Innovative Software Development in GIS

Innovative Software Development in GIS

Edited by
Bénédicte Bucher
Florence Le Ber

First published 2012 in Great Britain and the United States by ISTE Ltd and John Wiley & Sons, Inc.

Apart from any fair dealing for the purposes of research or private study, or criticism or review, as permitted under the Copyright, Designs and Patents Act 1988, this publication may only be reproduced, stored or transmitted, in any form or by any means, with the prior permission in writing of the publishers, or in the case of reprographic reproduction in accordance with the terms and licenses issued by the CLA. Enquiries concerning reproduction outside these terms should be sent to the publishers at the undermentioned address:

ISTE Ltd
27-37 St George's Road
London SW19 4EU
UK

www.iste.co.uk

John Wiley & Sons, Inc.
111 River Street
Hoboken, NJ 07030
USA

www.wiley.com

© ISTE Ltd 2012

The rights of Bénédicte Bucher and Florence Le Ber to be identified as the author of this work have been asserted by them in accordance with the Copyright, Designs and Patents Act 1988.

Library of Congress Cataloging-in-Publication Data

Innovative software development in GIS / edited by Florence Le Ber [and] Benedicte Bucher.
 p. cm.
Includes bibliographical references and index.
 ISBN 978-1-84821-364-7
 1. Geographic information systems. 2. Geography--Data processing. 3. Geomatics. I. Le Ber, Florence. II. Bucher, Bénédicte.
 G70.212.I556 2012
 910.285--dc23

2012008578

British Library Cataloguing-in-Publication Data
A CIP record for this book is available from the British Library
ISBN: 978-1-84821-364-7

Printed and bound in Great Britain by CPI Group (UK) Ltd., Croydon, Surrey CR0 4YY

Table of Contents

Chapter 1. Introduction 1
Bénédicte BUCHER and Florence LE BER

1.1. Geomatics software 2
   1.1.1. Digital geographical data 2
   1.1.2. GIS-tools 5
   1.1.3. Software innovation and geomatics
   research 9
1.2. Pooling 12
   1.2.1. The need for pooling and its relevance . . . 12
   1.2.2. Reflection opportunity on geomatics
   pooling 13
   1.2.3. Pooling within the MAGIS research group 15
1.3. Book outline 17
1.4. Bibliography 18

PART 1. SOFTWARE PRESENTATION 23

**Chapter 2. ORBISGIS: Geographical Information
System Designed by and for Research** 25
Erwan BOCHER and Gwendall PETIT

2.1. Introduction 25
2.2. Background history 26
2.3. Major functionalities 30

          2.3.1. Language and spatial analysis   30
          2.3.2. Representation: style and cartography . . .   35
          2.3.3. Other functionalities   36
               2.3.3.1. Visualization   36
               2.3.3.2. Editing   37
               2.3.3.3. OGC flux   38
     2.4. Architecture and graphical interface   39
          2.4.1. Architecture and models   39
               2.4.1.1. Creating a plugin   40
               2.4.1.2. Manipulating data   41
          2.4.2. Graphical interface   47
               2.4.2.1. The GeoCatalog   47
               2.4.2.2. The GeoCognition   47
               2.4.2.3. The Map and the TOC   48
     2.5. Examples of use   48
          2.5.1. Spatial diachronic analysis of
               urban sprawl   48
          2.5.2. Spatial hydrologic analysis   51
          2.5.3. Geolocation   56
               2.5.3.1. Geocoding   57
               2.5.3.2. Geographical rectification   57
     2.6. Community   61
     2.7. Conclusion and perspectives   63
     2.8. Acknowledgments   64
     2.9. Bibliography   64

**Chapter 3. GEOXYGENE: an Interoperable
Platform for Geographical Application
Development**   67
Éric GROSSO, Julien PERRET and Mickaël BRASEBIN

     3.1. Introduction   67
     3.2. Background history   68
     3.3. Major functionalities and examples of use . . .   69
          3.3.1. Generic functionalities   70
          3.3.2. Use case: building data manipulation . . .   70
               3.3.2.1. Data   70

             3.3.2.2. The data schema: the
                     `Building` class   72
             3.3.2.3. Object-relational mapping with OJB . .   73
             3.3.2.4. A processing example: building
                     urban areas   73
     3.4. Architecture 75
         3.4.1. The core 76
         3.4.2. First applicative layer: the basic
                applications   77
         3.4.3. Second applicative layer: the expert
                applications   78
             3.4.3.1. Semiology modules   80
             3.4.3.2. GEOXYGENE 3D module   80
             3.4.3.3. GEOXYGENE spatiotemporal
                     module 82
     3.5. Communities 84
     3.6. Conclusion 86
     3.7. Bibliography 88

**Chapter 4. Spatiotemporal Knowledge
Representation in AROM-ST**   91
Bogdan MOISUC, Alina MIRON, Marlène
VILLANOVA-OLIVIER and Jérôme GENSEL

     4.1. Introduction 91
     4.2. From AROM to AROM-ST   93
         4.2.1. AROM in context: a knowledge
                representation tool   93
         4.2.2. Originalities 95
         4.2.3. Why a spatiotemporal extension?   96
             4.2.3.1. Existence 96
             4.2.3.2. AROM's contribution   97
     4.3. AROM-ST   100
         4.3.1. Metamodel   100
         4.3.2. Objects and time relationships   102
         4.3.3. Space and time types   107
         4.3.4. Spatial modeling example with AROM . . .   108

4.4. From AROM-OWL to ONTOAST 112
4.5. Architecture 113
4.6. Community. 115
4.7. Conclusions and prospects 116
4.8. Bibliography 117

Chapter 5. GENGHIS: an Environment for the Generation of Spatiotemporal Visualization Interfaces 121
Paule-Annick DAVOINE, Bogdan MOISUC and
Jérôme GENSEL

5.1. Introduction 121
5.2. Context 122
   5.2.1. The SPHERE and SIDIRA applications: two applications devoted to visualizing data linked to natural risks 123
   5.2.2. GENGHIS: a generator of geovisualization applications devoted to multi-dimensional environmental data 125
5.3. Functionalities linked to the generation of geovisualization applications 127
   5.3.1. Use cases for GENGHIS 127
   5.3.2. Instancing the data model and the knowledge base 128
   5.3.3. Editing the presentation model 130
   5.3.4. Generating the geovisualization interface. 132
5.4. Functionalities of the geovisualization application generated by GENGHIS 133
   5.4.1. Spatial frame functionalities 135
   5.4.2. Temporal frame functionalities 135
   5.4.3. Informational frame functionalities 137
   5.4.4. Interactivity and synchronization principles 138
5.5. Architecture 140
5.6. Scope and user communities 141

   5.6.1. Natural risks: a privileged scope 141
     5.6.1.1. The SIHREN application 142
     5.6.1.2. The MOVISS application 144
   5.6.2. User community 146
  5.7. Conclusion and perspectives 147
  5.8. Acknowledgments 148
  5.9. Bibliography 149

Chapter 6. GEOLIS: a Logical Information System to Organize and Search Geo-Located Data 151
Olivier BEDEL, Sébastien FERRÉ and Olivier RIDOUX

  6.1. Introduction 151
  6.2. Background history 152
  6.3. Main functionalities and use cases 153
   6.3.1. Geographical data visualization and
      exploration 156
     6.3.1.1. Virtual layers: queries and
        extensions 157
     6.3.1.2. Visualizing a virtual layer: map and
        navigation index 158
     6.3.1.3. Building and transforming virtual
        layers: navigation links 163
   6.3.2. Representation of geographical data and
      spatial reasoning 168
     6.3.2.1. Representing spatial properties 169
     6.3.2.2. Representing spatial relations 172
   6.3.3. Use cases 174
     6.3.3.1. Direct search 175
     6.3.3.2. Targeted search 176
     6.3.3.3. Exploratory search 177
     6.3.3.4. Knowledge search 180
  6.4. Architecture 182
  6.5. Users and developers 184
  6.6. Conclusion 186
  6.7. Bibliography 186

Chapter 7. GENEXP-LANDSITES: a 2D Agricultural Landscape Generating Piece of Software 189
Florence LE BER and Jean-François MARI

7.1. Introduction 189
7.2. Context 190
7.3. Major functionalities 193
   7.3.1. Point generation 194
   7.3.2. Field pattern simulation 194
      7.3.2.1. Voronoï diagrams 195
      7.3.2.2. Random rectangular tesselation 196
   7.3.3. Cropping pattern simulation 198
      7.3.3.1. Stationary method 198
      7.3.3.2. Taking into account succession changes 199
      7.3.3.3. Future changes 199
   7.3.4. Post-production, spatial analysis, and formats 200
      7.3.4.1. Post-production 200
      7.3.4.2. Spatial analysis 200
      7.3.4.3. Formats, import, and export 201
7.4. Case uses 201
7.5. Architecture 204
   7.5.1. The application `Core` 205
   7.5.2. Separating graphical classes from business classes 205
   7.5.3. The plugin system 206
   7.5.4. Interface 206
7.6. Communities 207
7.7. Conclusion 209
7.8. Acknowledgments 209
7.9. Bibliography 210

Chapter 8. MDWEB: Cataloging and Locating Environmental Resources 215
Jean-Christophe DESCONNETS and Thérèse LIBOUREL

- 8.1. Introduction 215
- 8.2. Context 216
  - 8.2.1. Origins 216
  - 8.2.2. Positioning 218
- 8.3. Major functionalities and case uses 220
  - 8.3.1. Matching roles and functionalities 221
- 8.4. Cataloging functionality 224
  - 8.4.1. Notion of metadata 225
  - 8.4.2. Notion of metadata profile 226
  - 8.4.3. A simplified view of cataloging 228
  - 8.4.4. Cataloging in a multiuser context 232
  - 8.4.5. Cataloging extensions 234
    - 8.4.5.1. Help for metadata input 234
    - 8.4.5.2. Metadata exchange 236
- 8.5. Locating functionality 238
  - 8.5.1. Local and distant metadata querying 241
  - 8.5.2. Monolingual or multilingual querying 241
- 8.6. Administration functionality 244
- 8.7. Architecture 247
- 8.8. User community 249
- 8.9. Conclusion 251
- 8.10. Bibliography 253

Chapter 9. WEBGEN: Web Services to Share Cartographic Generalization Tools 257
Moritz NEUN, Nicolas REGNAULD and Robert WEIBEL

- 9.1. Introduction 257
- 9.2. Historical background 258
- 9.3. Major functionalities 262
  - 9.3.1. Uploading software tools 262
  - 9.3.2. Requesting a service 263
  - 9.3.3. Cataloging and discovering services 264

9.4. Area of use 265
   9.4.1. Usage 265
      9.4.1.1. Interactive mode 265
      9.4.1.2. Automatic mode 266
   9.4.2. User types 267
      9.4.2.1. Researchers 267
      9.4.2.2. Cartographic institutions (Institut Géographique National - IGN and others) 271
      9.4.2.3. GIS providers 271
9.5. Architecture 273
   9.5.1. WEBGEN services access 273
   9.5.2. A standard data model for generalization services 274
9.6. Associated communities 276
   9.6.1. Distribution 276
   9.6.2. Uses 276
   9.6.3. Contributors 276
9.7. Conclusion and outlook 277
9.8. Acknowledgments 279
9.9. Bibliography 279

PART 2. SUMMARY AND SUGGESTIONS 283

Chapter 10. Analysis of the Specificities of Software Development in Geomatics Research . . 285
Florence LE BER and Bénédicte BUCHER

10.1. Origin and motivations 286
   10.1.1. Targeted users and uses 286
   10.1.2. Motivations and foundations 287
10.2. Major functionalities, fields, and reusability . . 288
   10.2.1. Functionalities 288
   10.2.2. Fields 289
   10.2.3. Reusability 291

**Chapter 11. Challenges and Proposals for
Software Development Pooling in Geomatics** ... 293
Bénédicte BUCHER, Julien GAFFURI,
Florence LE BER and Thérèse LIBOUREL

    11.1. Requirements and challenges 294
        11.1.1. Pooling function implementations 294
            11.1.1.1. Reusing functions implemented in
                  geomatics 294
            11.1.1.2. The challenge of defining
                  interoperable interfaces 297
            11.1.1.3. The challenge of modular
                  development 299
        11.1.2. Pooling models and expertise 301
            11.1.2.1. The need for it 301
            11.1.2.2. A challenge: the diversity and gaps
                  in the existing expertise 302
    11.2. Solutions 303
        11.2.1. Reference frameworks and metadata ... 304
        11.2.2. Test cases to improve description of
            implemented functions and progress
            within a community 307
    11.3. Conclusion 311
    11.4. Bibliography 313

Glossary 317

List of Authors 325

Index 329

Chapter 1

Introduction

Research in geomatics must face major challenges to improve the management of the interaction of humankind with the planet at various levels. These challenges cover types of problems such as risk management (monitoring a volcano), sustainable development (the prevention of coastal erosion or the control of increasing urbanization in a given area), or even societal issues, such as the accompaniment and improvement of the integration of positioning techniques and their mobile applications in our everyday lives. To process these issues, we often need to turn to computers and develop software that can meet the requirements of the data handled. The goal of this book is to study the innovative software development activities carried out by geomatics research teams, and more specifically to analyze which of these development activities can be pooled, and whether it is relevant to do so, in the sense that it promotes research activities. We have chosen to focus on one aspect of geomatics research: the design of models and analysis methods to utilize geographical data.

Chapter written by Bénédicte BUCHER and Florence LE BER.

The rest of Chapter 1 clarifies the contextual elements that are essential to the study of geomatics, and more specifically the definitions of the terms used. We successively clarify the notions of geomatics software and pooling in our context before presenting the goals and structure of the book.

1.1. Geomatics software

Geomatics is a technical and scientific field derived from geography and computer science. It develops methods to represent, analyze, and simulate geographical space. Its goal is to improve the understanding of this space and the management of human activities and human interventions on the planet. Thus, the core activities of geomatics is made up of techniques of Earth observation as well as techniques of model design – mainly maps – useful for analysis and reasoning. The traditional spatial representations are printed maps, gazetteers, or lists of triangulation points. For the past 20 years, geographical data have become digital and geomatics has been characterized by the intensive use of computer science. This development is highlighted by two phenomena. The first is the increase in data, specifically satellite data, and this increase requires the development of automatic processing. The second phenomenon is the increasing role of geographical information in information infrastructures (use of maps on the Web, localized services, etc.).

1.1.1. *Digital geographical data*

A core specificity of geomatics is its data.

A primary aspect is the distance between the data and the information represented through them. This is partly due to the fact that space observation often happens through the measurement of physical signals that must then be interpreted into meaning. This distance between the data and the information is also due to the difficulty in representing the notion of position in space so as to carry out operations on the shapes of the objects and the spatial relations they represent. More specifically, a digital model of geographical space must render two important notions: positioning in space and the nature of the phenomena. Positioning in space is shown through projections, which relate the different parts of the Earth's surface to an ellipsoid linked to coordinates in a stable mathematical referential versus the Earth. Geographical projection is usually followed by a cartographic projection to view the data on a plane screen. Thus, part of the Earth's surface or its subsurface is positioned by a geometry provided with coordinates – eventually reduced to a point. From there, two major positioning methods exist: the *vector* and the lattice [COU 92]. For example, a road is generally represented by an object of linear geometry (corresponding to the axis of the road on the ground) with attributes taking its nature into account (identification number, classification, and type of surface). This is a *vector* model. However, in three-dimensional (3D) virtual worlds, roads are often not represented in the data as *vector* objects, but the human user can see them in the terrain image (due to texture). Other phenomena, such as air pressure, must be represented as fields which have a given value in any point of space. More specifically, discretized versions of these fields are used. These are lattice models. The continuous/discrete duality that exists at the level of the observed reality and in both models of representation can also be found in the principles of software development and sometimes leads researchers to adopt different approaches to study one phenomenon. When we study a city, for example, we use ORBISGIS with a preference for lattice representation manipulation and GEOXYGENE with a preference for the manipulation of *vector* objects. Overall, the choice of a representation often frames a domain of expertise and the joint manipulation of two types of representations remains complex even though there exist proposals to integrate them [LAU 00].

A second specificity of geographical data is the multiplicity of models built to represent geographical space in the data [BIS 97]. As [WOR 96] mentions it, geographical space isn't a *table top space*, which is a space observable from outside, similar to objects placed on a table. It is a space in which each person acts, and builds, a representation of the space in the context of his/her own action. For example, the information obtained from a geographical landscape isn't the same depending on whether the user is interested in road transport, risk management, or development. Differences appear at the level of the types of relevant objects: the watering places and pools are remembered by the fireman but not by the hauler. Differences also appear at the semantic and geometrical levels of detail: a building can be represented by its footprint and access points or in a simplified manner. Beyond the real-world ontology that is used – the categories of objects of the world observed and the logical diagram – the data also sometimes depend on specific rules of representation, such as a building of less than 20 m^2 is represented by an object of the `IsolatedConstruction` class if it is highly isolated (over 100 m from another building). Finally, the coding of the data and the required geometry discretization leads to other choices that can vary from one producer to the other.

All in all, the manipulation and interpretation of geographical data requires dedicated software and expertise. Moreover, the heterogeneities in the data stand in the way of pooling.

1.1.2. *GIS-tools*

A very popular type of software in geomatics is the geographical information systems tool (GIS-tool), which allows the manipulation of geographical data. The term "tool" allows us to distinguish the piece of software from the complete system made of data, software, and users. The term GIS generally refers to the entire system. From now on in this book, we will use the term GIS to refer to a GIS-tool. A GIS is characterized by many functionalities that are essential in geographical information and detailed as follows. Up until the 1990s, GIS software fulfilled all these functionalities. Monolithic architectures then became architectures made up of modules dedicated to various functionalities, which are required to use the geographical data. This evolution was helped by interface specifications between GIS components produced by International Organization for Standardization (ISO) and Open Geospatial Consortium (OGC)[1]. These specifications were deliberately made abstract at first so they wouldn't restrict the market. Implementations were quickly suggested and included into the standard ones: XML implementations for the interoperable Web service components and JAVA (GEOAPI) implementations for interoperable libraries. Today, the notion of GIS thus refers to an information system made up of data and functional modules. It holds definite interest for pooling since it encourages researchers to focus on their core interest and reuse functional modules for the supporting functions they need.

The GIS functionalities were referred to in France by the acronym "5A": "Acquire", "Afficher" ("Display"), "Archive", "Abstract", and "Analyze" [DEN 96]. A sixth "A", for

---

[1] The glossary presented at the end of the chapters gives an inventory of the organizations, tools, and formats quoted in this book.

"Anticipate", appeared along with the concern about sustainable development and simulation software.

The acquisition of geographical data in a GIS essentially consists of importing existing data. The software must thus be capable of reading the more common formats, which is greatly aided by the generalized adoption of standard formats such as ESRI's *shapefile* format or the GML format proposed by ISO/OGC [ISO 07]. The software must also allow the interpretation of models with imported data that is still problematic in spite of the many schema transformation tools such as the FME Workbench of the Safe Software company. Schema transformation is still an active research field today [BAL 07]. The software should also allow the direct creation or editing of geographical data, for example the description of a new piece of road by creating an object and drawing its geometry on a referential map. The function of integration and fusion mentioned by [STE 09] is also important at this stage. It is made difficult by the differences between the geographical space representations mentioned earlier. Indeed, a new list, which goes into more detail, of nine functionalities was recently suggested by [STE 09] to define a *GIS software* in a geographical encyclopedia: visualization, creation, editing, storing, integration/merger, transformation, query, analysis, and map writing. This list does not have acquisition but details the integration functionalities that are the key functions to build the database of a geographical information system. Finally, due to the rise of distributed architectures, the acquisition function is now doubled up with a function to discover existing data and existing functionalities. The MDWEB software presented in this book is a solution to this need provided by research teams (IRD and the University of Montpellier). The software was designed as a specific component of a GIS architecture, and turned out to be the most able to simply complete existing structures since it does not offer redundant structures and its interface is clearly identified.

The display is available in various functions: visualizing the data geometry, visualizing their attributes, and writing and visualizing a map from these data. The last function requires the association of geographical data and cartographic styles, and then to draw the corresponding figure, which means having graphical objects linked to geographical objects. The cartographic representation is specifically studied in the GENGHIS proposition described in this book. A cartographic style is the association between a piece of information and a graphical symbol. The styles are defined for object classes such as roads and avalanches and eventually refined within a class according to the attributes of the said class: roads, for example, are represented differently depending on the value of the "classification" attribute given to the road. It was for a long time impossible to transfer a legend (from the cartographic style definition) from one type of software to another, due to the lack of a standardized format. The current proposition of the OGC consortium, entitled *Styled Layer Descriptor*, aims to become just such a standard. Besides, within the context of pooling, display processing is not simply about being able to transfer a display specification from one type of GIS software to another. It is also about knowing how to adapt the display of data to the context. This issue has been studied in the field of collaborative GIS architectures, which aim to allow multiple actors (such as researchers) to work on the same set of data.

Abstraction corresponds to the possibility of creating and manipulating a more or less sophisticated model of geographical space. For example, if a user uploads a set of points from sensors, describing temperature and humidity data, a first level of abstraction would be to create zones in which these values are described as average and a second level of abstraction would be to create a classification of these zones. As we have mentioned it previously, there is no universal model to represent space. Within a GIS, abstraction also corresponds to the information formatting before its processing. There is also here a great diversity of abstraction models, which complementarity isn't always simple to explore, such as the abstractions based on agents or the abstractions based on cellular automata, such as [BAT 05] does for cities. The analysis carried out in a GIS corresponds to complex operations or reasoning on spatial properties or relations of the phenomena represented, as for example, the choice of the buildings surrounding an airport, or the calculation of an itinerary. In geographical information, the query is specifically complex since it often uses various criteria: the position in space, the nature, and the position in time. Moreover, the spatial criterion is multidimensional. Owing to their volume, it is usually necessary to index geographical data to allow these requirements. The construction of spatial indexes is made complex by the multidimensional nature of localization [KAM 08]. Moreover, the indexed objects can evolve, for example a fleet of taxis or planes [WOL 99]. Or the query itself can evolve, for example the query, made by a user on the move, for the closest Vélib bicycle docking stations in Paris, which is also called a continuous query [TER 92]. All this requires the organization of indexes so that they allow complex spatiotemporal queries, are not penalized by updates, and allow for a swift answer to a changing query. In this book, the GEOLIS software presents a different abstraction from the classical entity-relationship model to organize geographical data so that we can carry out exploration queries on them. Finally, the rise of the Web, and the first Web document, increased the importance of unstructured information searches. In this field, it is important to take into account the geographical dimension, since a major part of the queries made over the Web have a geographical dimension. Providing software that manages the spatial component in the indexation and the classification of answers improves search engine performance [PAL 10, PUR 07].

Analysis carried out in a GIS corresponds to the possibility of automatically carrying out complex operations or reasoning on the properties and spatial relations of the objects represented, such as the buildings around an airport, or the calculation of an itinerary. Among the functionalities defined by [STE 09], we have the query function. The query is specifically important and complex in geographical information for it requires the indexation of information under various crossed criteria: the position in space, the nature, and the position in time. In this book, the GEOLIS software offers a different abstraction from the classical entity-relationship model to organize these elements of geographical data aiming to make exploration queries on this data. The manipulation of spatiotemporal data has increased in importance, whether to manage moving objects or dynamic objects. The GENGHIS software presented in this book is dedicated to the implementation of spatiotemporal information systems (STIS).

1.1.3. *Software innovation and geomatics research*

Geomatics research aims to improve the knowledge and tools of geomatics, as well as promote the use of this knowledge and these tools and their integration into the information society. It is a multidisciplinary field, essentially made up of human and social science researchers and of computer science researchers, but also of researchers from other scientific fields such as law and signal processing. The research group MAGIS, "Méthodes et applications pour la géomatique et l'information spatial" (Methods and applications for geomatics and spatial information), covers 42 research laboratories and institutions. The research carried out in these laboratories focuses on localized services, new map types, models and applications for sustainable development, geographical information integration, spatial analysis, simulation, and geographical information science epistemology, among others.

Geomatics research is often inseparable from software usage to manipulate geographical data, whether they are complete GIS systems or specific modules. Researchers can be users. For example, geography researchers rely on GIS software to improve the knowledge of certain phenomena. Many models developed to study spatial phenomena, such as the erosion of agricultural land [DER 96], runoff and flooding [LAN 02], urban development [PIO 07, SIR 06], rely on sets of data stored in GIS that produce new data.

Researchers can also be developers, either to develop an *ad hoc* tool or suggest software innovations, which are developments whose scope is not restricted to solving a specific case. Some researchers work by developing extensions to existing software where these offer a programming interface, whether to offer new processing procedures or enrich a data model. These are typically works based on the ARCINFO software, widely used in American universities, or on the GRASS software, one of the first free pieces of GIS software. The ESRI international user conference thus welcomes some communications from researchers, the proof of which is the publication every year of a special issue of the scientific journal *Transactions in GIS* [WIL 10]. Other researchers ascribe to the development of a new tool. For example, this was the case for the graphical query interfaces CIGALES [MAI 90] or LVIS [BON 99], as well as for projects presented in this book.

Innovation can lie in the development of new analysis methods based on theories from mathematics or knowledge engineering fields. It can also be by suggesting a new interface to disseminate existing functionalities on a broader level.

Or yet, the innovation can be in the architecture itself. The range of corresponding software solutions is wide: 3D view reconstruction from pictures, multiagent architectures for distributed processing, a mobile data management system, robot cartographer, geographical search engine, etc. Innovation can also pertain to the development of tools specific to certain research programs, tools which allow the manipulation of geographical data, and which can be considered as future functionalities of GIS-tools. In this book, we will present GENEXP-LANDSITES a software dedicated to the simulation of virtual landscapes. It aims at exploring the variability of agricultural landscapes and considers different cases for the spatiotemporal organization of agricultural production. So GENEXP-LANDSITES belongs to the sixth "A" (Anticipate) of the GIS-tools. Let us emphasize that software innovation in geomatics is also due to other actors rather than researchers, such as the military or private companies. We can, for example, mention the GOOGLE MAPS API that offers a functionality for new users: integrating a map into a website with eventually a specific overlay. This functionality was already available through Web extensions for classic GIS software, but the innovation was to offer it to geomatics novices due to use of simple language.

Thus, change in geomatics is partly tied to the evolution in computer science, it follows them, and improves them. The main software innovations that have stood out in the field of geomatics in the last few years are in part the evolutions of architectures distributed toward the Web, *grid computing*, *cloud computing*, ubiquitous computer science, and ambient intelligence, as well as the phenomenon of the semantic Web, robotics, and miniaturization. In the last few years, for example, we find distributed GIS, especially on the Internet. These distributed architectures favor the implementation of participative GIS, which create new problems beyond the pooling of software components [MAR 08, TUR 08], due to the rise of ubiquitous environments, localized services and ubiquitous cartography that also rise in importance.

1.2. Pooling

The term "pooling" is derived from the verb "to pool", which can be defined as "to combine (as resources) in a common fund or effort" [MIS 93]. The term was used for information technology applications, as early as the introduction of these applications in small businesses and communities, to essentially mean the sharing of upkeep and update costs. The term "information technology pooling" is also used in research and training about data and resources, such as linguistic resources [PIE 08]: the goal is to offer access to all the information and knowledge produced by every person and thus promote knowledge dissemination and progress. In this book, we consider the term "pooling" as meaning the pooling of resources that come into play during the design and development of software, aiming for shared benefits. These resources can be varied: abstract models, code, programming interfaces, financing, or yet experience in project management.

1.2.1. *The need for pooling and its relevance*

The relevance of pooling is true for any field of research focusing on innovation. Indeed, a specific type of pooling is sharing methods, making one's methods accessible to others and vice versa. By sharing methods, we promote their improvements as well as the comparison between the methods, and thus progress. It also allows the pooling of effort on certain components, and thus enables us to go faster. This book holds such an example: the WEBGEN project aims to facilitate the comparison of different implementation with the same function of introduction, to facilitate the progression in this field of research. Another example of innovation pooling is the European project SPIRIT, whose goal is to design a search engine based on geographical knowledge. The design and implementation of the engine required the collaboration of teams specializing in research on information, spatial analysis, and visualization. The pooling of the software contributions of the various teams took place within a service-based architecture whose interface contracts were defined during a joint project [FIN 03].

We should also note that the research teams use and sometimes improve other pieces of software necessary to their activities in higher education and research in general, such as article writing, presentation preparation, sharing courses, setting up websites for conferences, as well as all the management activities required by an institution which relies on digital information systems. This book does not focus on these tools. That said, the necessity for pooling solutions to support these activities has been proved and an answer has actually been provided by the PLUME[2] project, or by the implementation of the university and higher education and research institution pooling agency[3]. Other initiatives focus on digital documents such as the HAL[4] or ARXIV[5] archive sites – which gather researchers' scientific publications – or even the ORI-OAI[6] software that creates digital document sharing portals between education and research institutions.

1.2.2. *Reflection opportunity on geomatics pooling*

A reflection on the possibilities of pooling software development projects carried out in geomatics research teams

---

2 http://www.projet-plume.org
3 http://www.amue.fr/
4 http://hal.archives-ouvertes.fr/
5 http://arxiv.org/
6 http://www.ori-oai.org/

is all the more timely now that the techniques allowing us to interoperate software components, to cooperate on the design of a module, to design reusable components, or even to reuse existing components have improved and are widespread in software development.

These techniques are first and foremost, in geomatics, norms and standards concerning interfaces between components manipulating geographical data. In the field of geomatics, these standards mostly come from the ISO and its technical committee TC211 as well as the OGC. Specification may concern exchanged data, as in the *Geographic Markup Language* norm for instance, or functionalities, as in the *Web Feature Service*, *Web Map Service*, and *Catalogue Service for the Web* norms.

These techniques also cover methods and correlated collaborative development tools, OMG method [OMG 08], software project management tools, such as *Enterprise Architect* as well as middleware techniques aiming to encourage the reuse of software components with mediation architectures or component architectures [KRA 06]. A key architecture is, for example, the Web service architecture that corresponds to an architecture based on loosely coupled components on a widely accessible network. Another proof of the maturity of middleware techniques is ubiquitous architectures [WEI 93, WAL 97].

A particularly interesting standard for us is the *Web Processing Services* standard proposed by OGC. It focuses on the online availability of geographical data processing to promote sharing and reuse.

Another element promoting pooling is the success of *open source* software projects. Indeed, having access to a software's sources promotes its understanding and reuse due to the code and debugging documentation.

Moreover, the new information and communication technologies promote the confrontation of disciplines around joint study objects (a societal phenomenon, a territory, a design project, etc.). We can mention the visualization breakthroughs which allow development experts, for example, to better communicate on their projects with experts of other disciplines (due to a virtual world representation). Let us also mention the technical breakthroughs in information integration, due to both the dissemination of spatial content aggregators (*mashups*) and the increasing adoption of techniques derived from artificial intelligence on the Web. We can then talk of pooling information and knowledge. This is one of the express purposes of the semantic Web [BER 01], and, for us here, more specifically of the geospatial semantic Web [LIE 06]. Achieving this goal starts first and foremost with an effort to describe the information (in standard XML/RDF formats) available on the Web. This also requires the development of ontologies (for which we have the standard language OWL [DEA 04]) and automatic reasoning mechanisms which allow us to interpret the information described. The AROM and AROM-ST extensions we will describe in this book are a step in this direction.

1.2.3. *Pooling within the* MAGIS *research group*

This study was carried out within the "Exchange, Pooling, Design" project of the MAGIS research group, and of its predecessor SIGMA. This reflection welcomed contributions from external researchers when they provided a new point of view, useful to the reflection. The WEBGEN work, which has previously been mentioned, falls into this category.

The research group has four research axes or poles: the "Sensor" pole, the "Model" pole, the "Analysis" pole, and the "Decision" pole. We will now outline how each of these poles functions within geomatics research and how pooling – in the sense we have used here – is required for each of these research axes:

– The "Sensor" pole deals with the sources of geographical data acquisition and communication means. The tools developed are not only aimed at capturing data, but can also adapt to the user's needs. More specifically, the development of GPS satellite localization means and the improvement in precision enable us not only to pinpoint static objects but more and more to follow moving objects, including individuals, to which we can then offer various services.

– The "Model" pole focuses on various research components, from the perception in a geographical environment of phenomena of all shapes (thematic diversity), scales, and spatial or temporal granularities to their digital representation. The developed models are meant, on the one hand, to formalize concrete and abstract concepts linked to geographical objects or processes in space, and on the other hand to take into account various perceptive modalities: the verbal and textual forms of description, the visual, the naive geography, etc. These new forms of geographical environment description create various issues (interoperability and integration of the design with the usual representation forms of geographical information).

– The "Analysis" pole deals with an old and fundamental field of geographical information research, which is still very much relevant today due to the very rapid increase in the volume of available data and the need to have tools and diversified and renewed methods to interpret them. One of the current problems is the integration of multisource data; another is the visual restitution of data, which requires the implementation of numerous geographical concepts that have yet to be identified and clarified.

– The "Decision" pole focuses on the mobilization of geographical information within the frame of a decision process. These processes, personal or collective, public or private, are carried out by heterogeneous and multiple actors, by users and providers of information. We must thus understand the use and the production of geographical information by the various actors, local authorities or environmental agencies, commercial and industrial businesses, etc. The questions of use, organization, appropriation, and communications must be asked within a renewed frame, always attentive to emerging practices.

All these axes focus on different aspects which bring us back to the issue of pooling: production and dissemination of data, integration and interoperability of modes, integration of various data and expertise sources, etc. The whole set proves the need to share data, models, and knowledge. A first – and fairly advanced – possibility is to implement norms enabling communication between different types of software. The other possibilities are examined in this book through the description of different research or software development experiments.

1.3. Book outline

Chapters 2 to 9 of the book aim to give a more detailed analysis of the reasons for which geomatics researchers are led to develop software solutions. They describe different specific development experiments using a common backdrop that helps by comparing the experiments and makes the book easier to read. This backdrop was defined jointly by all the authors of the chapters describing software development projects. Its specifications are as follows:

– short introduction;

– history: scientific and technical context of development, rationality, founding principles, and project management;

– major functionalities and how-to: basic functionalities and expert functionalities;

– architecture: interface types for possible reuse;

– associated communities: carriers, contributors, dissemination, and effectiveness of prospective use;

– conclusion: feedback from experiments, perspectives, and legal considerations;

– bibliography.

Following these detailed presentations, we will sketch an innovative GIS software development case "cartography". We offer typologies to describe these software developments according to their different characteristics (the goals they aim for, the contexts, functions, data, interfaces, users, expertises, etc.). We will analyze the needs and obstacles to pooling.

Based on this analysis, we will then present proposals to improve pooling in software developments carried out by geomatics research teams.

1.4. Bibliography

[BAL 07] BALLEY S., Aide à la restructuration de données géographiques sur le Web – Vers la diffusion à la carte d'information géographique, PhD in computer science, University of Paris-Est Marne-la-Vallee, 2007.

[BAT 05] BATTY M., *Cities and Complexity*, The Massachusetts Institute of Technonology Press, Cambridge, MA, 2005.

[BER 01] BERNERS-LEE T., HENDLER J., LASSILA O., "The semantic web", *Scientific American*, vol. 1, pp. 34–43, 2001.

[BIS 97] BISHR Y., Semantic aspects of interoperable GIS, PhD Thesis, ITC, Enschede, The Netherlands, 1997.

[BON 99] BONHOMME C., TRÉPIED C., AUFAURE M.-A., LAURINI R., "A visual language for querying spatiotemporal databases", *Proceedings of the 7th International Symposium on Advances in Geographic Information Systems – ACM-GIS' 1999*, Kansas City, USA, pp. 34–39, 1999.

[COU 92] COUCLELIS H., "People manipulate objects (but cultivate fields): beyond the raster-vector debate in GIS", *Proceedings of the International Conference GIS – From Space to Territory: Theories and Methods of Spatio-Temporal Reasoning in Geographic Space*, vol. 639 of LNCS, Pisa, Italy, 1992.

[DEA 04] DEAN M., SCHREIBER G., BECHHOFER S., VAN HARMELEN F., HENDLER J., HORROCKS I., MCGUINESS D., PATEL-SCHNEIDER P., STEIN L., OWL Web Ontology Language – Reference, W3C Recommendation, World Wide Web Consortium, 2004.

[DEN 96] DENÈGRE J., SALGÉ F., *Les systèmes d'information géographique*, Que sais-je?, PUF, Paris, 1996.

[DER 96] DE ROO A.P.J., WESSELING C.G., RITSEMA C.J., "LISEM: a single-event physically based hydrological and soil erosion model for drainage basins", *Hydrological Processes*, vol. 10, no. 8, pp. 1107–1117, 1996.

[FIN 03] FINCH D., Specification of system functionality, deliverable D4 no. 1101, SPIRIT Technical Group (IST-2001-35047), 2003.

[ISO 07] ISO TC211, ISO 19136 – Geographic Information – Geographic Markup Language (GML), Report, ISO International Standard, 2007.

[KAM 08] KAMEL I., "Indexing, Hilbert R-tree, spatial indexing, multimedia indexing", *Encyclopedia of GIS*, SpringerScience/Business Media, New York, pp. 507–512, 2008.

[KRA 06] KRAKOWIAK S., Intergiciel et construction d'applications réparties, Ecole d'été ICAR, 2006.

[LAN 02] LANGLOIS P., DELAHAYE D., "RuiCells, automate cellulaire pour la simulation du ruissellement de surface", *Revue Internationale de Géomatique*, vol. 12, no. 4, pp. 461–487, 2002.

[LAU 00] LAURINI R., GORDILLO S., "Field orientation for continuous spatio-temporal phenomena", *Proceedings of the International Workshop on Emerging Technologies for Geo-Based Applications*, Ascona, Switzerland, 2000.

[LIE 06] LIEBERMAN J., Geospatial semantic web interoperability experiment report, Report, Open Geospatial Consortium Inc., 2006.

[MAI 90] MAINGUENAUD M., PORTIER M.-A., "Cigales: A graphical query language for geographical information systems", *Proceedings of 4th International Symposium on Spatial Data Handling*, Zurich, Switzerland, pp. 393–404, 1990.

[MAR 08] MARTIGNAC C., TEYSSIER A., THINON P., CHEYLAN J.-P., "SIG participatifs et développement: contributions de l'expérience de la réforme foncière malgache", *International Conference on Spatial Analysis and Geomatics – SAGEO' 2008*, Montpellier, France, 2008.

[MIS 93] MISCH F.C., (ed.), *Merriam-Webster's Collegiate Dictionary*, Merriam-Webster, Incorporated, Springfield, MA, 1993.

[OMG 08] OMG, Software & Systems Process Engineering Meta-Model Specification, v2.0, 2008.

[PAL 10] PALACIO D., CABANAC G., SALLABERRY C., HUBERT G., "Measuring geographic IR systems effectiveness in digital libraries: evaluation framework and case study", *Proceedings of the 14th European Conference on Research and Advanced Technology for Digital Libraries – ECDL'10*, Glasgow, Scotland, pp. 340–351, 2010.

[PIE 08] PIERREL J.-M., "De la nécessité et de l'intérêt d'une mutualisation informatique des connaissances sur le lexique de notre langue", *Congrès Mondial de Linguistique Française, Paris*, French Institute of Linguistics, 2008.

[PIO 07] PIOMBINI A., FOLTÊTE J.-C., "Evaluer les choix d'itinéraires pédestres en milieu urbain", *Revue Internationale de Géomatique*, vol. 17, pp. 207–225, 2007.

[PUR 07] PURVES R., CLOUGH P., JONES C., ARAMPATZIS A., BUCHER B., FINCH D., FU G., JOHO H., KHIRINI A., VAID S., YANG B., "The design and implementation of SPIRIT: a spatially-aware search engine for information retrieval on the Internet", *International Journal of Geographic Information Systems (IJGIS)*, vol. 21, no. 7, pp. 717–745, 2007.

[SIR 06] SIRET D., MUSY M., RAMOS F., GROLEAU D., JOANNE P., "Développement et mise en œuvre d'un SIG 3D environnemental urbain", *Revue Internationale de Géomatique*, vol. 16, no. 1, pp. 71–91, 2006.

[STE 09] STEINIGER S., WEIBEL R., "GIS Software – a description in 1000 words", *Encyclopeadia of Geography*, SAGE Publication, London, UK, 2009.

[TER 92] TERRY D.B., GOLDBERG D., NICHOLS D., OKI B.M., "Continuous queries over append-only databases", *Proceedings of the SIGMOD*, 1992.

[TUR 08] TURKUCU A., ROCHE S., "Classification fonctionnelle des public participation GIS", *Revue Internationale de Géomatique*, vol. 18, no. 4, pp. 429–442, 2008.

[WAL 97] WALDO J., WYANT G., WOLLRATH A., KENDALL S., "A note on distributed computing", *Mobile Object Systems: Towards the Programmable Internet*, LNCS 122, Springer Verlag, 1997.

[WEI 93] WEISER M., "Some computer science issues in ubiquitous computing", *Communications of the ACM*, vol. 36, no. 7, 1993.

[WIL 10] WILSON J.P., "GIScience research at the Thirtieth Annual ESRI International User Conference", *Transactions in GIS*, vol. 14, no. 1, 2010.

[WOL 99] WOLFSON O., SISTLA P., CHAMBERLAIN S., YESHA Y., "Updating and querying databases that track mobile units", *Distributed and Parallel Databases Journal (DAPD)*, vol. 7, no. 3, pp. 257–288, 1999, special issue on Mobile Data Management and Applications.

[WOR 96] WORBOYS M.F., "Metrics and topologies for geographic space", *Advances in GIS Research II, Proceedings of 7th International Symposium on Spatial Data Handling*, Taylor and Francis, Delft, The Netherlands, pp. 365–375, 1996.

Part 1
Software Presentation

Chapter 2

ORBISGIS: Geographical Information System Designed by and for Research

2.1. Introduction

ORBISGIS[1] is a geographical information system (GIS) dedicated to scientific modeling and experimenting. ORBISGIS has been developed at the IRSTV, a French research institute dedicated to urban science and techniques[2] since April 2007, within the "urban data" Federative Research Project (FRP) framework whose goal is to provide methods and tools to gasp the challenges of urban environments.

There are three main objectives within the "urban data" FRP:

– data acquisition techniques (teledetection, model reconstruction, on-site measurements, etc.);

Chapter written by Erwan BOCHER and Gwendall PETIT.
1 http://www.orbisgis.org/, accessed September 2011.
2 http://www.irstv.fr/, accessed September 2011.

– representation and processing of spatial information (storage, modeling, multiscale simulation: time + 3D);

– sharing geographical information.

ORBISGIS was built on top of free and *open source* libraries. It is distributed under a GPL 3 license (*open source*)[3]. ORBISGIS's goal is to be a federating tool, gathering within the research units of the IRSTV all the methods and processed data linked to geographical information, irrespective of the research field they come from (sociology, civil engineering, urban architecture, geography, economy, environment, etc.).

This chapter is divided into five sections in which we describe the background history of ORBISGIS (section 2.2), present its major functionalities (section 2.3), detail its architecture (section 2.4); present three use cases (section 2.5), and end by giving a few elements of information about the developer and user community (section 2.6).

2.2. Background history

IRSTV is an FR CNRS 2488 research federation and a federative structure of the French Ministry for Higher Education and Research. IRSTV is made up of 15 laboratories and carries out interdisciplinary research in the fields of modeling and sustainable urban management [HÉG 06]. Its research activities are focused around three major themes:

– an interdisciplinary urban observation system (urban teledetection, and multidisciplinary experimentations site – MWS);

---

3 http://gplv3.fsf.org/, accessed September 2011.

– an integrated environmental modeling of the city (integrated urban microclimatology, sound atmospheres, urban data modeling, and GIS);

– governance, design, and sustainable urban management.

This multidisciplinary aspect is the cause of a great disparity in the use of geographical information, whether it is the data (storing and modeling), the tools used to exploit it, or the processing chains implemented [BOC 07a, BOC 08a, BOC 08b]. The diversity in GIS software is twofold: a diversity in storage support and a formal diversity in the description of data. It leads to a division of geographical knowledge, which is, in a way, the opposite of IRSTV's goals: to develop an integrated vision of all the urban physical phenomena, methods, tools, and actor systems which contribute to the sustainable management of the city.

To overcome these gaps and reinforce a federative spirit, the outlines of a GIS for urban modeling and management appear within the framework of the regional program MeigeVille – "Modélisation environnementale intégrée et gestion durable de la ville" – which stands for urban integrated modeling and sustainable management [HÉG 06]. The goal of this GIS is to design the theoretical and instrumental bases of a capitalization tool of urban environment knowledge as well as analysis methods and management techniques [HÉG 06]. It is at this point that the plan to create a GIS platform to ensure coordination and animation was set in motion.

In December 2006, the GIS platform was put in the hands of a research engineer specializing in spatial reference data. The platform is structured around the development of two platforms:

– a spatial data infrastructure (SDI);

– a community GIS.

28   Innovative Software Development in GIS

By referring itself to geographical data sharing and exchange best practices, which are written down in national and international recommendation documents [CLI 94, NEB 04, INS 07], the GIS platform lays down the bases of an interoperable architecture made of (Figure 2.1) [BOC 07c]:

– a data repository to store information;

– a third application, called Geoservices, to share data using OGC standards;

– a Web cartographic portal to view, explore and search data;

– a GIS software, called OrbisGIS, to view, process, display and push data.

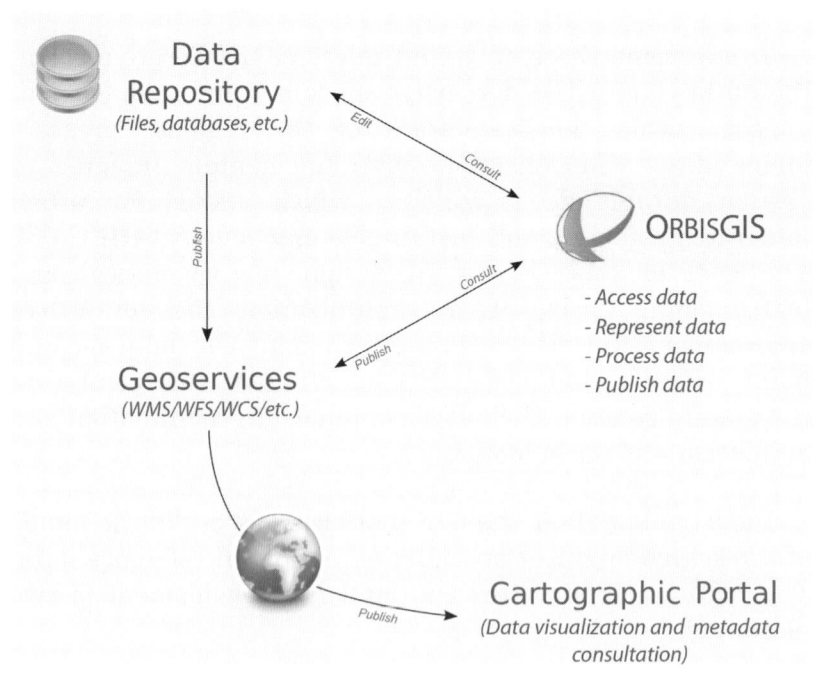

Figure 2.1. ORBISGIS *within the SDI project at the IRSTV*

As the first part of the SDI puzzle, ORBISGIS was developed to answer the requirements of research. Indeed, it is during the implementation of a chain of analyses and processing of urban soil tenure within the MeigeVille project [BOC 08b] that gaps of formalism and interoperability between GIS-tools were highlighted with regard to the manipulation of geographical objects. Each tool had its own language, concepts, and terms to describe a geographical process, and it was consequently very difficult to exchange methods unless mediators were developed for each tool.

In this situation there appeared the idea of an advanced language, able to access the main geographical data formats and structures (*vector* or *raster*) while respecting the international standards as much as possible. Relying on the *Simple Features* SQL (SFS) norm [HER 06a, HER 06b], this is the main processing language of the ORBISGIS platform. It enabled all the IRSTV researchers to build a common library of processes working for issues, such as spatial hydrology, urban tissue evolution analysis, and noise mapping (French National Research Agency projects such as AVUPUR, EvalPDU, and VEGDUD[4]). Within this context, a new and more federative approach to geographical information came to life in the IRSTV, leading researchers to build a GIS together which would be dedicated to the analysis of urban environments.

The first *beta* version of ORBISGIS was released at the end of June 2007, during the 8th Libre Software Meeting [BOC 07b]. Since then, the following versions have been released:

---

4 AVUPUR: *assessing the vulnerability of peri urban rivers*; EvalPDU: evaluation of the environmental impacts of a plan or urban shifts and their socioeconomic consequences; and VEGDUD: the role of plants in sustainable urban development.

- 3.0 (Barcelona): February 2011;
- 2.2.0 (Paris): December 2009;
- 2.1.0 (Vienna): June 2009;
- 2.0.0 (Ostrava): January 2009;
- 1.2.0 (Naoned): August 2008;
- 1.1.0 (Boston): July 2008;
- 1.0.0 (Girona): June 2008.

2.3. Major functionalities

One of the major particularities of ORBISGIS is its Generic Datasource Management System (GDMS) library [BOC 08b, LED 09]. Beyond data access, this library enables us, through a specific query language derived from SQL, to process geographical (*vector* and *raster*) data and allocated data, using on the one hand a set of functions in accordance with the SFS specifications of the Open Geospatial Consortium (OGC) and on the other hand the specific functions developed for research needs. ORBISGIS is the graphical interface designed to explore and represent the data manipulated by GDMS.

2.3.1. *Language and spatial analysis*

The issue of a generic language to manipulate the geographical data and carry out spatial analysis is not new. As early as the end of the 1970s, [TOM 79] suggested a set of conventions and operators to manipulate *georasters*, which are *raster* images with metadata related to a geographical position. This formalism, called MAP ALGEBRA, was used in many GIS such as the SPATIAL ANALYST[5] for the ESRI ARCGIS© or GRASS's *raster* analysis module[6].

At the beginning of 2000, the OGC, with the SFS specification, defined a set of operators and spatial predicates as well as a syntax to manipulate, in SQL, *vector* geometries of types such as `Point`, `Linestring`, `Polygon`, etc. [HER 99, HER 06a, HER 06b]. This document formalizes the many attempts presented by the scientific community, such as those made by [EGE 88b, EGE 88a, EGE 89] or [GOH 89, GUT 88], and for the first time endorses a language dedicated to *vector* geographical data. This specification was welcomed by consensus, especially by private companies which did not wait long to adopt it. The POSTGIS[7] spatial add-on for the relational database POSTGRESQL is a perfect illustration of this. In 2010, the website logged, on average, 800 downloads of the source code a month.

However, the SFS standard only provides us a partial answer to manipulate geographical data. The operators and predicates are only defined for *vector* data. *Raster* data processing and the topological reasoning (graph routing, arithmetic on *raster* graphics or convolution, etc.) are no longer listed as OGC improvement priorities. Yet in a multidisciplinary context like the context at the IRSTV, where the city is observed at various scales and where studied objects are varied (urban tasks, air pollution, and surface hydrology), research works require data sources as well as varied models and structures.

We can provide a simple example, the spatial analysis of urban area evolution. Starting with a set of satellite

---

5 http://www.esri.com, accessed September 2011.
6 http://grass.fbk.eu/, accessed September 2011.
7 http://postgis.refractions.net/, accessed September 2011.

images taken on various dates, the user can extract sets of pixels corresponding to urban objects (buildings, parking lots, etc.). These sets are transformed into a collection of polygons (vectorization) which describe the urban space. They are stored in layers separated by their observation date. If the scales, for the operator in teledetection, are set at the level of the object identification method, for the geographer or the developer, the focus will be the study of the phenomenon's distribution. Thus, the data from the image classifications will be aggregated, combined with *vector* databases which represent the different levels of land administration (local authorities, major cities, and their suburbs).

In the current situation, which is not specific to the IRSTV, there are two work approaches to implement this processing chain:

– either it is entirely carried out by the teledetection operator who only provides the geographer with the results;

– or the geographer is provided with the image results, and he/she then analyzes them with his/her own tools.

In this context, one of the major challenges is to promote the dissemination of not only data, but also methods. The goal is to decompartmentalize analyses and spatial processes by making them independent of the software in which they are carried out, and to promote exchanges.

With that in mind, at the beginning of 2007, the IRSTV started a reflection on the use of data processing languages. The conclusions led to the development of the GDMS library. This library integrates an application that analyzes and carries out SQL instructions on *vector* or *raster* geographical data as well as on alphanumerical data. The SQL grammar is based on the SQL-92 specifications and the *Simple Features* SQL standard. However, for reasons previously mentioned, the grammar was extended to work on the *raster* data. This ability to manage the *raster* format is also present in the commercial software ORACLE SPATIAL with the GEORASTER module.

With the GDMS library, it is thus possible to produce a buffer zone on a file in *shapefile* format. In that case, two instructions are carried out:

– With the `Register` function, the user first saves the `myShapefileFile.shp` file, gives it a name (`myShapeFile`), and accesses the data:

```
SELECT Register('/tmp/myShapeFileFile.shp',
        'myShapeFile').
```

– The user then applies a buffer zone of 20 m on the geometries of the `monShapeFile` file, which are found in the `the_geom` column:

```
SELECT ST_Buffer(the_geom, 20) as the_geom
        FROM myShapeFile.
```

This query is equivalent to the query used in the POSTGRESQL SGBD, with its POSTGIS spatial extension.

However, GDMS is different when it comes to *raster* format manipulation. In the following example, we will dissect an image from a geometry due to the function `ST_CropRaster`:

```
SELECT ST_CropRaster(a.raster,b.the_geom)
        as raster
    FROM myRaster a, myShapeFile b;
```

Let us note that first, the user will have declared the file was *raster* to access it:

```
SELECT Register('/tmp/myRasterFile.tif',
        'myRaster');
```

The result produces a new *raster* image whose content is stored in a *raster*-type column. The GDMS data model will then be presented.

Figure 2.2 presents the major processing categories which are supported by SQL language today. All in all, GDMS has almost 200 functions. Among the most common are the SFS specification spatial operators and predicates (ST_Intersects, ST_Difference, ST_Buffer, ST_Contains, ST_Touches, etc.) as well as advanced functions:

– ST_D8Direction, ST_D8Accumulation, ST_Watershed for hydrology;

– ST_CreateGrid, ST_Compacity for morphological analysis;

– ST_Delaunay, ST_ConstrainedDelaunay for triangulation.

VECTOR FUNCTIONS
- Quality control
- Conversion
- Generalization
- Spatial operators and predicates
- Topology (network and plane)
- Triangulation

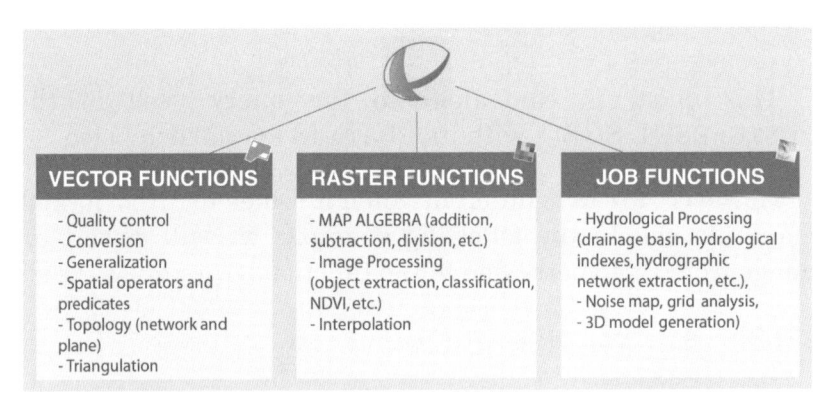

RASTER FUNCTIONS
- MAP ALGEBRA (addition, subtraction, division, etc.)
- Image Processing (object extraction, classification, NDVI, etc.)
- Interpolation

JOB FUNCTIONS
- Hydrological Processing (drainage basin, hydrological indexes, hydrographic network extraction, etc.),
- Noise map, grid analysis,
- 3D model generation)

Figure 2.2. ORBISGIS's *main spatial analysis functions*

The reflection carried out during the definition of the SQL grammar and the development of GDMS allowed us to highlight the gaps present in the SFS norm. Provided bases to query *vector* geometries, SFS must today evolve to cover other essential uses in spatial analysis. With GDMS-R [LED 09], we have proved that it is possible to manipulate *raster* graphics in spatial SQL. This demonstration is the basis on which we will rely on to formalize *raster* operators and suggest an evolution in the SFS specification to OGC.

2.3.2. *Representation: style and cartography*

Like most GIS, ORBISGIS can affect graphic styles and build cartographies. Figure 2.3 summarizes the main available methods. From a simple classification by unique value to the creation of a map with confronted symbols, ORBISGIS offers an almost complete functional coverage to represent geographical data.

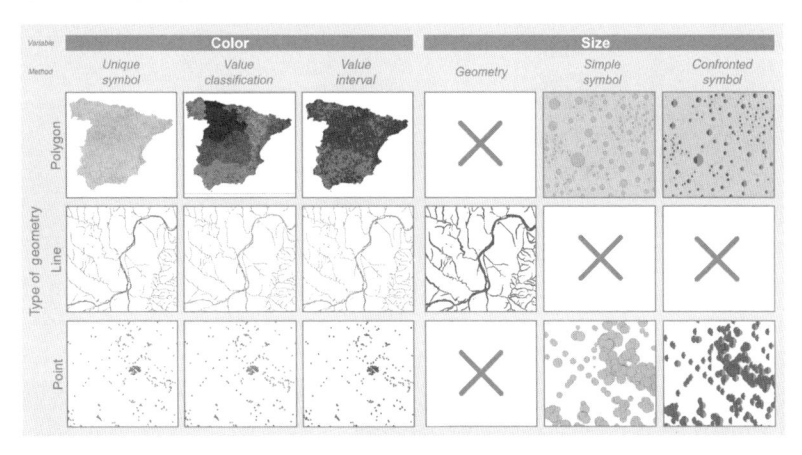

Figure 2.3. *Style and cartography. For a color version of this figure, see www.iste.co.uk/Bucher/innovgis.zip*

Style editing and legend creation are created by calling up the "Edit Legend" interface on a layer selected in the TOC (*Table of Contents*). The "Edit Legend" window is divided into three parts (Figure 2.4). The leftmost part lists the legends that are chosen by the user. The choice of a representation method is made in the "Legend Selection" window, activated by clicking on the "+" icon in the legend list toolbar. When a legend is validated by the user, a graphical interface comes up with its own methods (background color, symbol type, display scale, etc.).

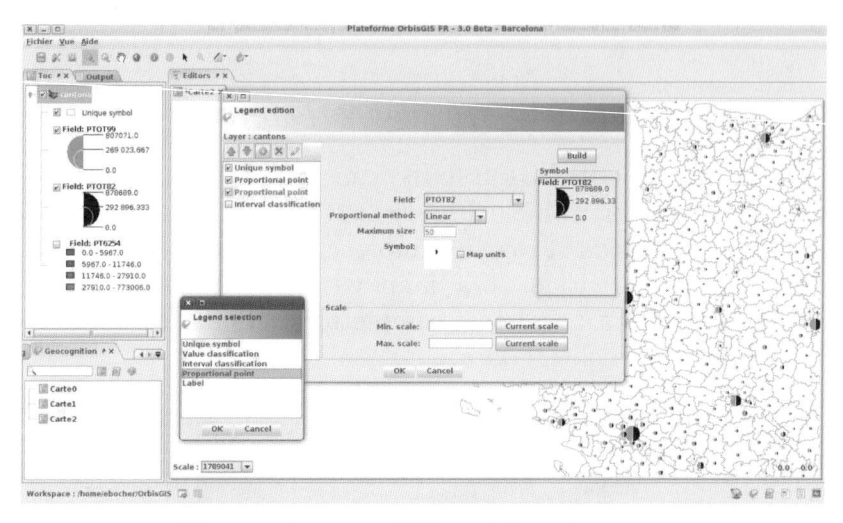

Figure 2.4. ORBISGIS's *multilegend system*

Let us also specify that the model used to manage the legends by layer is plural. Indeed, a same layer in ORBISGIS can have various cartographic representations: proportional symbols, unique symbols, etc. Each of these legends was activated – inactivated in the TOC. An example of this is given in Figure 2.4.

2.3.3. *Other functionalities*

As other GISs, ORBISGIS has classical management, visualization, and editing functions for geographical data.

2.3.3.1. *Visualization*

Geographical data visualization is carried out in the "Map" window. It has navigation functions (zoom, pan) as well as interrogation and measurement tools (surface, length, and angle measurement). The user can also display a table's attributes in a window. This window will offer tools to produce statistics in a numerical field, change the name of a column, and sort the values of a column in an ascending or a descending order, select rows, etc.

2.3.3.2. *Editing*

Depending on the data format, ORBISGIS enables us to edit geometries as well as attributes. Table 2.1 lists the *vector* and alphanumerical formats currently supported by the platform[8].

Type of data	Storage	Name	Reading/Writing
Vector	File	ESRI *shapefile*	Reading & Writing
		GDMS	Reading & Writing
		MIF/MID	Reading
		DXF	Reading
		SOLENE	Reading & Writing
		VRML	Reading & Writing
	Relational database	POSTGRESQL/POSTGIS	Reading & Writing
		H2 spatial	Reading & Writing
Alphanumerical	File	DBF	Reading & Writing
		CSV	Reading & Writing
	Relational database	POSTGRESQL	Reading & Writing
		H2	Reading & Writing
		HSQLDB	Reading & Writing

Table 2.1. ORBISGIS-*supported formats*

Geometry editing is carried out in the "Map" view in a layer selected in the TOC. The available functions are:

---

8 SOLENE is an insolation, illumination, and thermal radiation simulation software developed by the CERMA Laboratory.

38 Innovative Software Development in GIS

– creating points, lines, and polygons, as well as multigeometries; a geometry figure can, for example, have more than one polygon;

– deleting geometry;

– modifying, moving, adding, and deleting a geometry's peaks.

As for attributes, ORBISGIS has an advanced interface that offers the possibility for the user to not only add a column in a table depending on a listed type, but also to specify constraints in this column. The creation of a geometry field can, for example, be limited to a Point type and a 2D or 2.5D dimension (Figure 2.5).

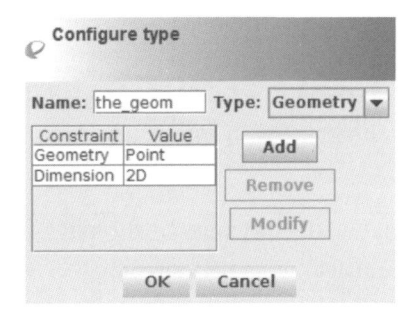

Figure 2.5. *Creating a field with constraints*

2.3.3.3. *OGC flux*

ORBISGIS only supports the Web Map Service (WMS) protocol, versions 1.1–1.3. The recovery of a layer from a WMS server is carried out through Geocatalog. The recovered WMS image is interpreted as a DataSource object in the GDMS data model. It is then added in the map view to be visualized (Figure 2.6).

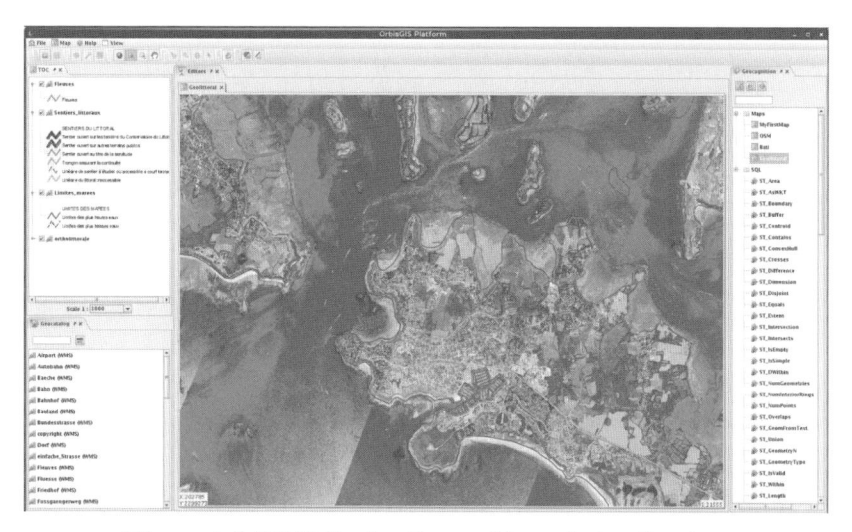

Figure 2.6. *WMS flux loading and legend visualization*

2.4. Architecture and graphical interface

The ORBISGIS platform was designed with a modularity in mind, both at the level of the architecture and the graphical interface. Let us present these two aspects.

2.4.1. *Architecture and models*

The ORBISGIS application is organized around two libraries:

– the GDMS library for *vector* and *raster* data access and processing;

– the ORBISGIS-Core library made up of the graphical interfaces and "man-machine" actions, a layer model that allows us to manage the data as well as their legend and the *plugin* system.

The ORBISGIS architecture is summarized in Figure 2.7.

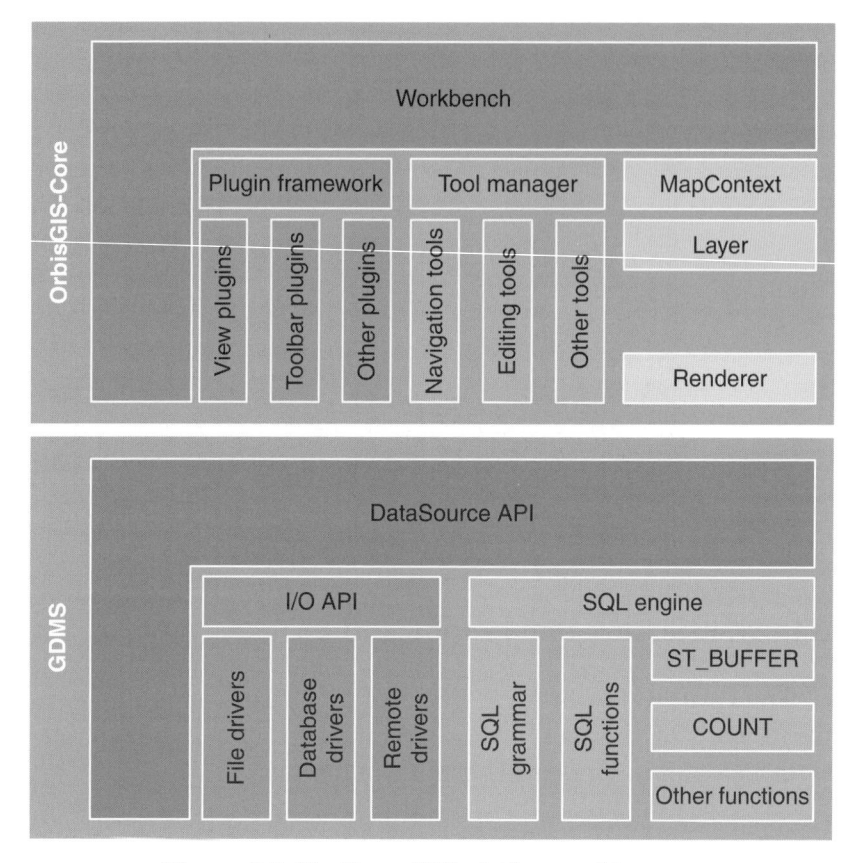

Figure 2.7. *The* ORBISGIS *platform architecture*

2.4.1.1. *Creating a plugin*

Each component of the ORBISGIS interface is a *plugin* that can itself be used by another *plugin*. The code written down in Table 2.2 illustrates an example of *plugin* writing when the goal is to count the number of objects in a geographical layer. The *plugin*'s operating procedure is commented on in the code.

The *plugin* can then be called upon in two ways:

– it is either delivered with the core application in which case it is declared to be in the `OrbisConfiguration` class;

– or the *plugin* is externalized in a JAVA (.jar) archive. In that case, the developer must write a specific Extension class which will become the interface. This class will initialize one or many *plugins* which will be in charge of launching the platform through the *plugin* manager. The compiled .jar file must be placed in the ext directory located in the folder containing the compiled ORBISGIS libraries.

2.4.1.2. *Manipulating data*

Geographical data manipulation in ORBISGIS is carried out through the use of two application programming interfaces (API):

– the GDMS API;

– the ORBISGIS-Core LAYERMODEL API.

2.4.1.2.1. The GDMS API

The GDMS API is an abstraction layer that reads, writes, and processes geographical and alphanumerical objects. The data stored in files or databases are encapsulated during the Source object declaration. If the Source object is a unique reference to a physical piece of data, it can, however, also point to a virtual piece of data. In that case, the view aggregates the data from the two sources.

To access the piece of data, GDMS builds a DataSource object. This object is an interface on the GDMS *drivers* layer. The pieces of data are built in a "row/column" tabular model equivalent to an SGBDR model, in this case a table (Figure 2.8). A DataSource object contains a collection of values and a metadata schema. A value that corresponds to a DataSource cell attribute can be of various kinds: String for text, Integer for integers, and Geometry for geometry. The Metadata object contains a list of fields with their constraints. A geometry-type field can, for example, be limited to points.

42 Innovative Software Development in GIS

```
public class CountFeaturesTOCMenu extends AbstractPlugIn {
   /**
    * Method specifying the type of plugin to be built.
    * We are dealing here with a ''Count objects'' menu
which will appear in the TOC when the user right-clicks **/
   public void initialize(PlugInContext context) throws Exception
   {
     WorkbenchContext wbContext = context.getWorkbenchContext();
     WorkbenchFrame frame =
             wbContext.getWorkbench().getFrame().getToc();
     context.getFeatureInstaller().addPopupMenuItem(frame, this,
         new String[] { "Count objects" }, wbContext);
   }
   /**
    * Method enabling us to specify the display conditions of
the ''Count objects'' menu when right-clicking on a TOC layer **/
   public boolean is Enabled()  {
     // The running MapContext is retrieved.
It provides information
     // on the number of layers loaded, etc...
     MapContext mc = getPlugInContext().getMapContext();
     if (mc != null)  {
     // If the number of layers is equal to 1 then
     // the plugin is visible.
         if (mc.getSelectedLayers().length == 1)  {
           return true;     }    }
     return false;  }
   /**
    * Plugin execution method **/
   public boolean execute(PlugInContext context) throws Exception
   {
     MapContext mapContext = getPlugInContext().getMapContext();
     ILayer[] selectedLayers = mapContext.getSelectedLayers();
     // The GDMS object containing the layer data is retrieved.
     SpatialDataSourceDecorator sds = selectedLayers[0]
         .getSpatialDataSource();
     // The result is displayed in ORBISGIS with
     // the OutputManager Window.
     Services.getOutputManager().print(
         "Number of objects: " + sds.getRowCount(), Color.RED);
     return false;  }    }
```

Table 2.2. *Example of a* plugin *code in* ORBISGIS

The reader will note in Figure 2.8 that `DataSource` supports two types of geographical data: `Geometry` and `Raster`.

– The `Geometry` object is a vectorial representation (`Point, Line, Polygon`) in the shape of a series of coordinates in (`X,Y,Z`). To manipulate geometries, GDMS uses the JAVA Topology Suite library[9]. The latter establishes the *Simple Features* SQL norm to describe the coordinates in `WKT` (*Well-Known Text*) or `WKB` (*Well-Known Binary*) and suggests spatial operators and predicates (`Intersection, Union, Difference`, etc.).

– The `Raster` object is an interface of the `GeoRaster` object, which is itself an overlay encompassing the `ImagePlus` object of the IMAGEJ library[10] as well as georeferencing metadata. Like the JAVA Topology Suite library, the IMAGEJ library is used to carry out processing on the image(s) contained in the `GeoRaster` object: convolution, thresholding, object extraction, etc.

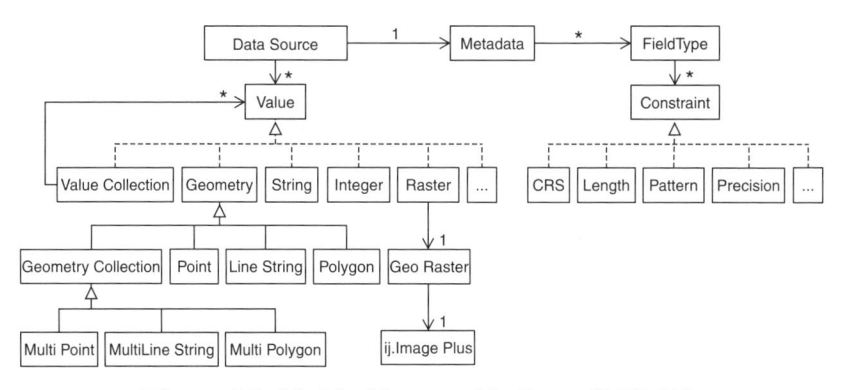

Figure 2.8. *Model of data used by* GDMS *[LED 09]*

The `DataSource` object is manipulated in the following way:

– `getRowCount()` gives the number of rows.

---

9 http://tsusiatsoftware.net/jts/main.html, accessed September 2011.
10 http://rsbweb.nih.gov/ij/, accessed September 2011.

- `getFieldNames()` gives the names of the fields in a table.

- `getFieldIndexByName(String fieldName)` gives the position of a field according to its name.

- `getRow(long rowIndex)` gives the table's data line at the rowIndex index as a table.

- `insertEmptyRow()` gives a new empty row.

- `insertFilledRow(Value[] values)` adds a new line containing values.

- `insertEmptyRowAt(long rowIndex)` adds a new line according to the position.

- `insertFilledRowAt (long rowIndex, Value[] values)` adds a new line with values according to the position.

- `deleteRow(long rowIndex)` deletes a line of data.

- `getFloat (long rowIndex, int fieldIndex)`, `getLong (long rowIndex, int fieldIndex)`, `getInt(long rowIndex, int fieldIndex)`, etc., give the value of a field on a given line of data.

Let us mention that to interact with the [DataSource] object, the developer must rely on three methods:

- `open()`: open the *driver* linked to the data source. This can be authorized in reading and writing mode.

- `commit()`: save the modifications on the data source.

- `close()`: close the *driver*.

Outside of the API, the developer can use SQL language to manipulate data. In that case, he/she will call up the DataSourceFactory. In the following example, a 20 m buffer zone was created around prairie-like parcels.

```
DataSourceFactory dsf = new DataSourceFactory();
dsf.executeSQL("SELECT ST_BUFFER(the_geom, 20)
   FROM parcel WHERE type = 'prairie';");
```

The instruction execution is delegated to GDMS's internal SQL engine that analyzes syntax and builds the operator chain necessary to the query. In SQL, the developer takes advantage of the functions available in GDMS (`ST_Buffer`, `ST_Area`, `StringToDouble`, etc.). However, he/she can also extend the SQL syntax by adding his/her own functions. In that case, he/she writes a new class implementing the `Function` interface. This class must then be saved in the function manager to be recognized by GDMS [BOC 08b].

2.4.1.2.2. The LAYERMODEL API

The LAYERMODEL API is the link between the physical piece of data interpreted by GDMS and its graphical representation. It formalizes the concept of geographical layer that corresponds in our case to a data source and a set of legends.

A legend is a collection of symbols. Five classes of "primitive" symbols are provided by the API: the `PointSymbol`, the `LineSymbol`, the `PolygonSymbol`, the `LabelSymbol`, and the `ImageSymbol`. The developer then expands on them to design more complex symbols.

For example, the `CirclePointSymbol` class, which consists of drawing a circle of a given size around a point, expands the `PointSymbol` class by overloading the `size` variable builder. The `PointSymbol` class has information such as filling color, thickness, and the color of the outline or the rendering unit. The `draw` method of the `PointSymbol` class will then have to be adapted to draw the circle according to the parameters specified in the builder.

46   Innovative Software Development in GIS

```
public class CirclePointSymbol extends PointSymbol {
   int size = 1;
   CirclePointSymbol(Color outline, int lineWidth, Color
     fillColor, int size, boolean mapUnits) {
       super(outline, lineWidth, fillColor, mapUnits);
       this.size = size;
  }}
```

To end this presentation of LAYERMODEL, we will comment the next pseudo-code which allows us to load a layer of polygons in ORBISGIS. Polygons are represented with a red filling color, a three-pixel thickness and a yellow outlining color.

```
// Retrieve ORBISGIS's source management and create a
// new layer containing a shapefile and a
// default legend
ILayer layer = getDataManager().createLayer
    (new File("src/test/resources/data/bv_sap.shp"));

// Create a legend containing a single symbol
UniqueSymbolLegend legend
        = LegendFactory.createUniqueSymbolLegend();

// Drawing variables
Color fillColor = Color.RED;
int lineWidth = 3;
Color outlineColor = Color.YELLOW;

// Create the adapted symbol
Symbol polygonSymbol = SymbolFactory.createPolygonSymbol
(outlineColor, lineWidth, fillColor);

// Assign the symbol to the legend
legend.setSymbol(polygonSymbol);

// Assign the legend to the layer
layer.setLegend(legend);

// Assign the layer to the LayerModel
getLayerModel().addLayer(layer);
```

2.4.2. *Graphical interface*

ORBISGIS is an entirely modular multiscreen application. All its windows can be added, deleted, resized, and dissociated from the general interface. The user is the one who shapes the interface to his needs.

However, for a basic operating ORBISGIS, the user must manipulate the four components (Figure 2.9) presented.

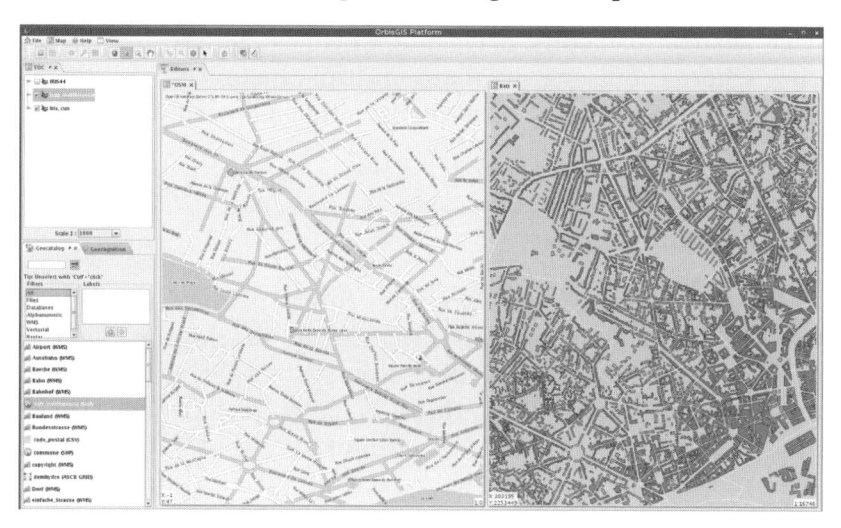

Figure 2.9. *Display of two "map" views, as well as the basic* ORBISGIS *components:* `TOC`, `GeoCatalog`, `GeoCognition`

2.4.2.1. *The GeoCatalog*

The `GeoCatalog` object is the window allowing us to connect, in ORBISGIS, geographical and alphanumerical data that will be displayed in the "map" or attribute windows.

2.4.2.2. *The GeoCognition*

The `GeoCognition` object is a file manager. It can create "map" windows. Each file can be organized, and moved in directories created by the user. The `GeoCognition` also contains all the available functions in the GDMS library.

2.4.2.3. *The Map and the TOC*

The "map" view is the visualization window of geographical data. It allows us to navigate, zoom, and edit objects. The TOC corresponds to the layer manager. It allows us to order the display of layers and access the functions of a layer by right-clicking on it. This is the case, for example, when the user wishes to modify a legend. However, unlike the "map" view where the user can create multiple instances, the TOC is a single window whose content is refreshed according to the active map.

2.5. Examples of use

In this section, we present three examples of how ORBISGIS is used. The first is a diachronic spatial analysis of urban sprawl, the second is a spatial hydrologic analysis, and finally the third is data geolocation. These applications are far from exhaustive, but they show the platform's possibilities, ranging from the use of spatial operators on *vector* data or the use of mixed data in SQL (*raster* and *vector*) to the design of a query chain to geolocate the data of an address at a parcel level.

2.5.1. *Spatial diachronic analysis of urban sprawl*

Urban sprawl is characterized by an evolution of urbanized surfaces on the ground, to the detriment of agricultural parcels, wooded zones, and natural ecosystems. Its study implies a comparison in time and over the same spatial unit.

Our example will focus on the Nantes metropolitan area. We will apply a set of SQL instructions whose final goal is to produce a synthesis map translating the evolution (gain–loss) of urbanized areas. Table 2.3 lists the data used as well as the names of the layers (tables) used in the SQL queries.

Data	Geometry	Layer Name
Zoning database (Z-DB), produced by the Loire-Atlantique local authorities for the years 1999 and 2004	Polygon and multipolygons	- zbd99 - zbd04
Towns in the Nantes metropolis	Polygon	- com_nantes

Table 2.3. *List of data used*

The processing chain, executed in the ORBISGIS console, can be divided into four major steps:

– selecting urbanized surface and dividing the Nantes metropolis area for 1999 and 2004;

– unifying the urbanized surface by town for 1999 and 2004;

– calculating the urbanized surfaces and density for 1999 and 2004;

– comparing the two years (1999 and 2004) and calculating the surface and density difference.

The script executed in the SQL console is presented in Tables 2.4 and 2.5.

After the processing, we can observe that except for a single town west of Nantes, the urbanized surfaces have increased over the metropolitan area (Figure 2.10).

```sql
-- A- Extract the urbanized surface (C_NIVEAU =1) by town
-- For 2004
CREATE TABLE Z04_urb_nm as SELECT ST_INTERSECTION
(a.the_geom, b.the_geom) AS the_geom, a.CODE_INSEE
FROM com_nantes a, zdb04 b
WHERE ST_INTERSECTS(a.the_geom, b.the_geom)
AND b.C_NIVEAU=1;
-- For 1999
CREATE TABLE Z99_urb_NM AS SELECT ST_INTERSECTION
(a.the_geom, b.the_geom) AS the_geom, a.CODE_INSEE
FROM com_nantes a, zdb99 b
WHERE ST_INTERSECTS(a.the_geom, b.the_geom)
AND b.C_LEVEL=1;

-- B- Calculate the urbanized surface in each town
-- For 2004
CREATE TABLE stats_04 AS SELECT SUM(ST_Area(the_geom)) AS
sum_surf04, CODE_INSEE FROM Z04_urb_NM GROUP BY CODE_INSEE;
-- For 1999
CREATE TABLE stats_99 AS SELECT SUM(ST_Area(the_geom)) AS
sum_surf99, CODE_INSEE FROM Z99_urb_NM GROUP BY CODE_INSEE;

-- C- Join the town table and the surface table
-- For 2004
CREATE TABLE join_04 AS SELECT a.*, b.sum_surf04
FROM com_nantes a, stats_04 b
WHERE a.CODE_INSEE = b.CODE_INSEE;
-- For 1999
CREATE TABLE join_99 AS SELECT a.*, b.sum_surf99
FROM com_nantes a, stats_99 b
WHERE a.CODE_INSEE = b.CODE_INSEE;

-- D- Calculate the urbanized surface density (in %)
    per town
-- For 2004
ALTER TABLE join_04 ADD COLUMN density_urb04 NUMERIC;
UPDATE join_04 SET density_urb04
=((sum_surf04/ST_AREA(the_geom))*100);
-- For 1999
ALTER TABLE join_99 ADD COLUMN density_urb99 NUMERIC;
UPDATE join_99 SET density_urb99
=((sum_surf99/ST_AREA(the_geom))*100);
```

Table 2.4. SQL *script of the analysis of urbanized surface evolution (part 1)*

```
-- E- Calculate the evolution of the urbanized
surface between 2004 and
-- 1999. Create a new layer which represents the
--   urbanized surface variation
CREATE TABLE Evol_Z_urban AS SELECT a.*except sum_surf99,
density_urb99, (b.sum_surf04 - a.sum_surf99) AS diff_area
FROM join_99 a, join_04 b WHERE a.CODE_INSEE=b.CODE_INSEE;
-- Add a new field called "diff_density" representing
-- the difference in density
ALTER TABLE Evol_Z_urban ADD COLUMN diff_density NUMERIC;
-- Update this field
UPDATE Evol_Z_urban SET diff_density=
((diff_area/ST_AREA(the_geom))*100);
```

Table 2.5. SQL *script of the analysis of the urbanized surface evolution (part 2)*

2.5.2. *Spatial hydrologic analysis*

Spatial hydrology consists of using geographical data to spatially describe physical phenomena: erosion, water movement, etc. A digital terrain model (DTM) of a *raster* type is often used to produce morphological (slope, orientation, etc.) or hydrologic (drainage basin limit, dendritic drainage systems, etc.) indicators. These indicators, coupled with data such as land use and human development localization, enable us to bring elements of understanding to the hydrological behavior of a terrain (identification of zones where surface runoff is concentrated, impact of agriculture on the ground, and erosion mechanisms). In the following example, we will present a use of ORBISGIS to localize the evolution of urbanized spaces in a drainage basin around a theoretical hydrographic network. This hydrographic network is extracted from a DTM *raster*. A 200 m buffer zone is applied around the hydrographic network to limit the spatial analysis. The approach is summarized in Figure 2.11. The land use data are identical to the data used previously. The DTM

52   Innovative Software Development in GIS

raster is produced by triangulating contour lines of SCAN 25 (IGN, the French National Geographic Institute). It covers the peri-urban drainage basin of the Chézine, northwest of the Nantes city.

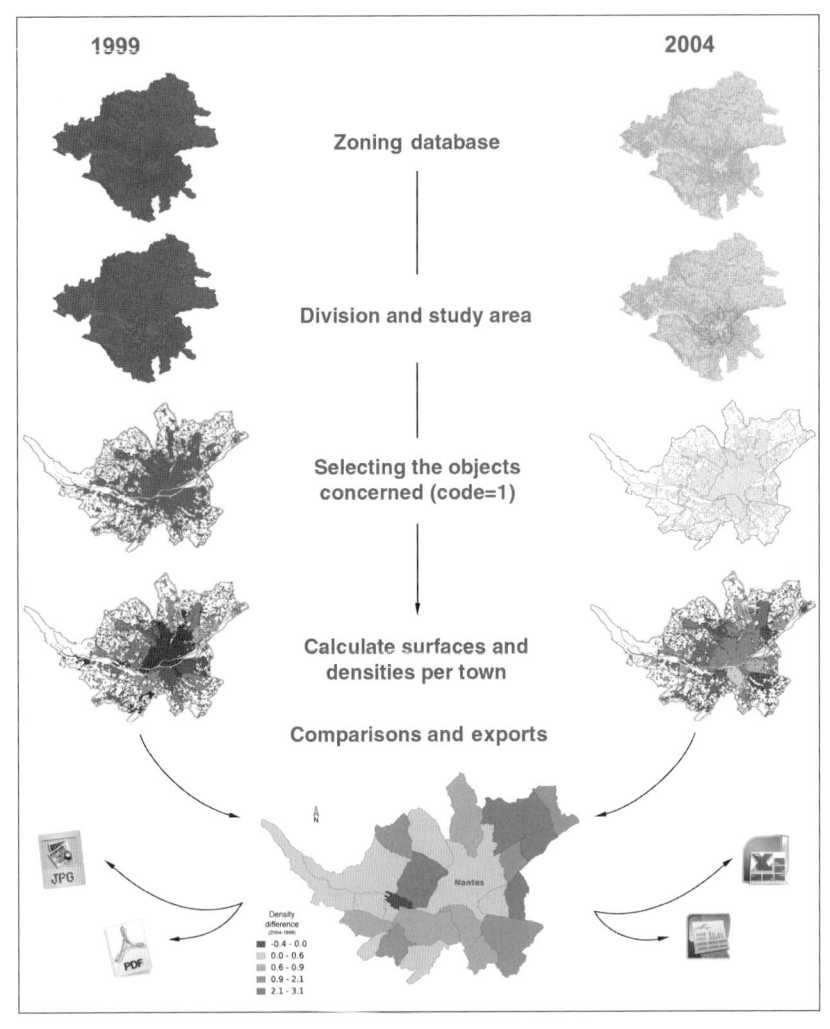

Figure 2.10. *Diagram of the analysis of the urbanized surface evolution*

ORBISGIS 53

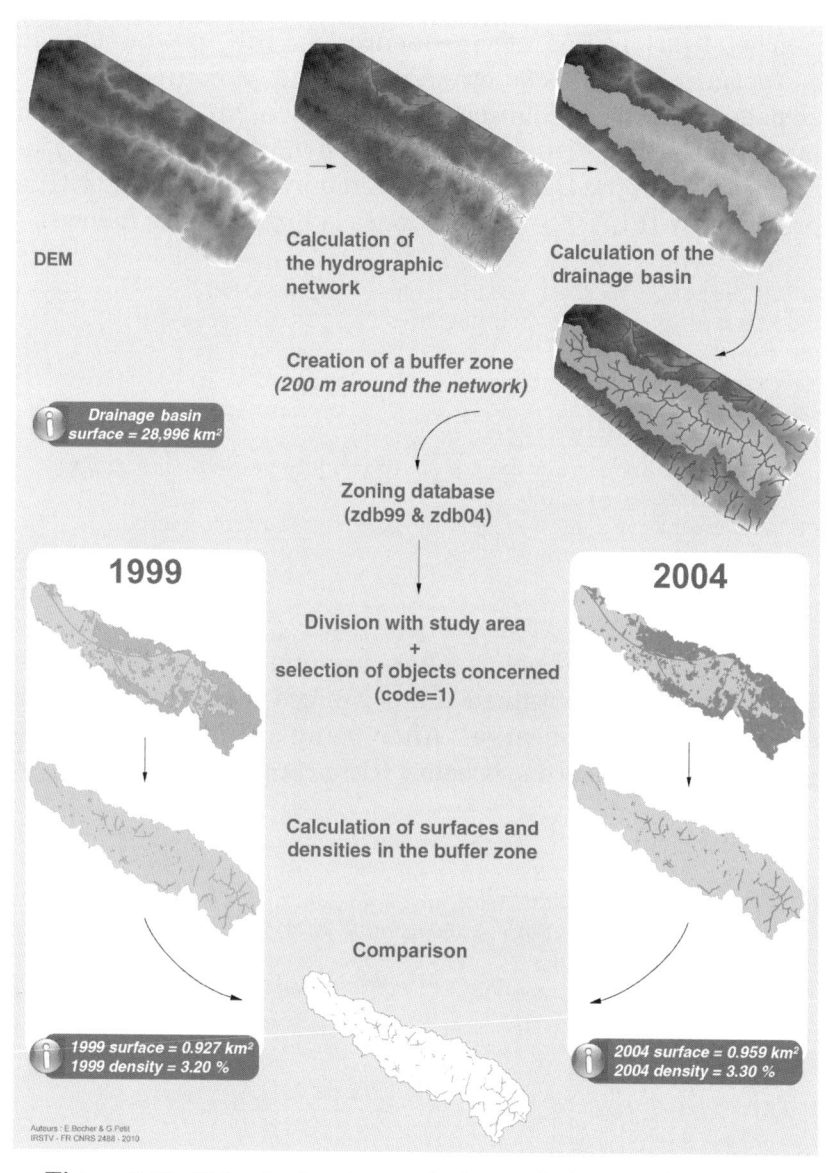

Figure 2.11. *Urbanized surface evolution analysis method around a theoretical hydrographic network. For a color version of this figure, see www.iste.co.uk / Bucher / innovgis.zip*

The first processing sequence is the production of a topographical pixel orientation grid according to D8 strongest slope calculation method, then an accumulation lattice [OCA 84]. To guarantee the hydrologic continuity of surface runoff, the DTM is first filtered by using the algorithm proposed by [PLA 02]. The sequence is presented as follows:

```
-- Filtering the DTM to fill depressions
CREATE TABLE filled AS SELECT ST_FillSinks(raster, 0.1)
AS raster FROM DEM;
-- Creating an orientation grid for pixels
CREATE TABLE dir AS SELECT ST_D8Direction(raster)
AS raster FROM filled;
-- Creating an accumulation latticed based on the pixel
-- orientation grid
CREATE TABLE acc AS SELECT ST_D8Accumulation(raster)
AS raster FROM dir;
```

Following this, a theoretical hydrographic network is calculated by linked pixels whose accumulation value is over 1,500. The pixels obtained are organized along the Strahler stream order classification. After being vectorized, only the geometries whose order is below 6 are retained:

```
-- Extract organize pixels whose accumulation
-- value is over 1500
CREATE TABLE strahler AS SELECT ST_D8StrahlerStreamOrder
    (d.raster, a.raster, 1500) FROM dir d, acc a;
-- Vectorize pixels
CREATE TABLE allrivers AS SELECT ST_VectorizeLine()
    FROM strahler;
-- Filter the geometries whose Strahler stream order
-- is below 6.
CREATE TABLE rivers_low_6 AS SELECT * FROM allrivers
    WHERE order<6;
```

A 200 m buffer zone is then created:

```
CREATE TABLE buffer200 AS SELECT ST_BUFFER(the_geom, 200)
    AS the_geom FROM rivers_low_6;
```

In parallel, after identification of the geographical coordinates of its outlet, the drainage basin of the Chézine is extracted:

```
CREATE TABLE basin AS SELECT ST_D8Watershed(raster,
 GeomFromText('POINT (303172.520884 2254429.11725)'))
    AS the_geom
FROM dir;
```

The final step consists of dividing land use modes for the years 1999 (z99) and 2004 (z04) in the limit of the drainage basin. Only the geometries whose C_Level field values are equal to 1 are kept. This value corresponds to urbanized areas.

```
-- For 1999
CREATE TABLE z9_urb_bv AS SELECT ST_INTERSECTION
 (a.the_geom, b.the_geom) AS the_geom, a.*EXCEPT the_geom
 FROM z99 a, basin  b WHERE ST_INTERSECTS(a.the_geom,
 b.the_geom) AND a.C_LEVEL=1;
-- For 2004
CREATE TABLE z04_urb_bv AS SELECT ST_INTERSECTION
 (a.the_geom, b.the_geom) AS the_geom, a.*EXCEPT the_geom
 FROM z04 a, basin  b WHERE ST_INTERSECTS(a.the_geom,
 b.the_geom) AND a.C_LEVEL=1;
```

The reader will note that the instructions hold the keyword EXCEPT. This keyword was introduced into the ORBISGIS SQL language to facilitate column filtering. So if the user wishes to select fields a, b, c, d,..., i, j in a table without selecting fields g and h, he/she will write:

– either   SELECT a, b, c, d, e, f, i, j FROM my_layer;

– or SELECT *{EXCEPT g, h} FROM my_layer.

With the keyword EXCEPT, column filtering is simplified. We not only avoid rewriting a list of fields, which can sometimes be long, but also avoid any possibility of typographical error.

For the next to last processing sequence, the land use per drainage basin is divided by the 200 m buffer zone around the theoretical hydrographic network:

```
-- For 1999
CREATE TABLE z99_urb_buff AS SELECT ST_INTERSECTION
 (a.the_geom, b.the_geom) AS the_geom, a.*EXCEPT the_geom
 FROM z99_urb_bv a, buffer200 b
 WHERE ST_INTERSECTS(a.the_geom, b.the_geom);
-- For 2004
CREATE TABLE z04_urb_buff AS SELECT ST_INTERSECTION
 (a.the_geom, b.the_geom) AS the_geom, a.*EXCEPT the_geom
 FROM z04_urb_bv a, buffer200 b
 WHERE ST_INTERSECTS(a.the_geom, b.the_geom);
```

Finally, the following script allows us to calculate urbanized surfaces and their density in connection to the drainage basin area. Let us note that the urbanized surfaces less than 200 m from the theoretical hydrographic network have barely increased between 1999 and 2004 (Figure 2.11):

```
-- For 1999
CREATE TABLE z99_hydro_surf AS SELECT
 ST_UNION(a.the_geom) as the_geom,
 SUM(ST_AREA(a.the_geom)) AS surf_urb99,
 ((SUM(ST_AREA(a.the_geom))/SUM(ST_AREA(b.the_geom)))*100)
 AS density_urb99 FROM z99_urb_buff a, basin b;
-- For 2004
CREATE TABLE z04_hydro_surf AS SELECT
 ST_UNION(a.the_geom) as the_geom,
 SUM(ST_AREA(a.the_geom)) AS surf_urb04,
 ((SUM(ST_AREA(a.the_geom))/SUM(ST_AREA(b.the_geom)))*100)
 AS density_urb04 FROM z04_urb_buff a, lim_bv_region b;
```

2.5.3. *Geolocation*

The following example concerns the issue of geolocating data at the address. We wish to link real estate transactions carried out by notaries in the years 2000 and 2008 in the city of Nantes to a parcel entity of the computerized parcel cadaster (CPC) provided by the French Tax Directorate (TD). The transaction data come from the Perval database[11].

Each recording in this base has 100 characteristics, including the street name, zip code, town or city name, as well as the cadastral section code and parcel code (TD).

However, in more than 85% of cases, the parcel code isn't mentioned. It is thus impossible to carry out a direct comparison between the transaction and its location.

The approach we suggest taking is divided into two major stages: a geocoding stage and a geographical rectification stage.

2.5.3.1. *Geocoding*

Geocoding is linking a Perval address with an address reference system for which each X and Y coordinate is known. To this end, we use Google's geocoder[12]. The resulting file is loaded into the ORBISGIS platform and transformed into a set of points with the following instruction:

```
CREATE TABLE pervalGeographic SELECT ST_MakePoint(X, Y) AS
   the_geom, * FROM pervalGeocode;
```

The result is a new geographical layer called pervalGeographic.

2.5.3.2. *Geographical rectification*

The goal of rectification is to "re-locate" each recording of the pervalGeographic layer on a CPC parcel. The methodology is summarized in Figure 2.12.

---

11 http://www.perval.fr/, accessed September 2011.
12 http://www.batchgeocodeur.mapjmz.com/, accessed September 2011.

58 Innovative Software Development in GIS

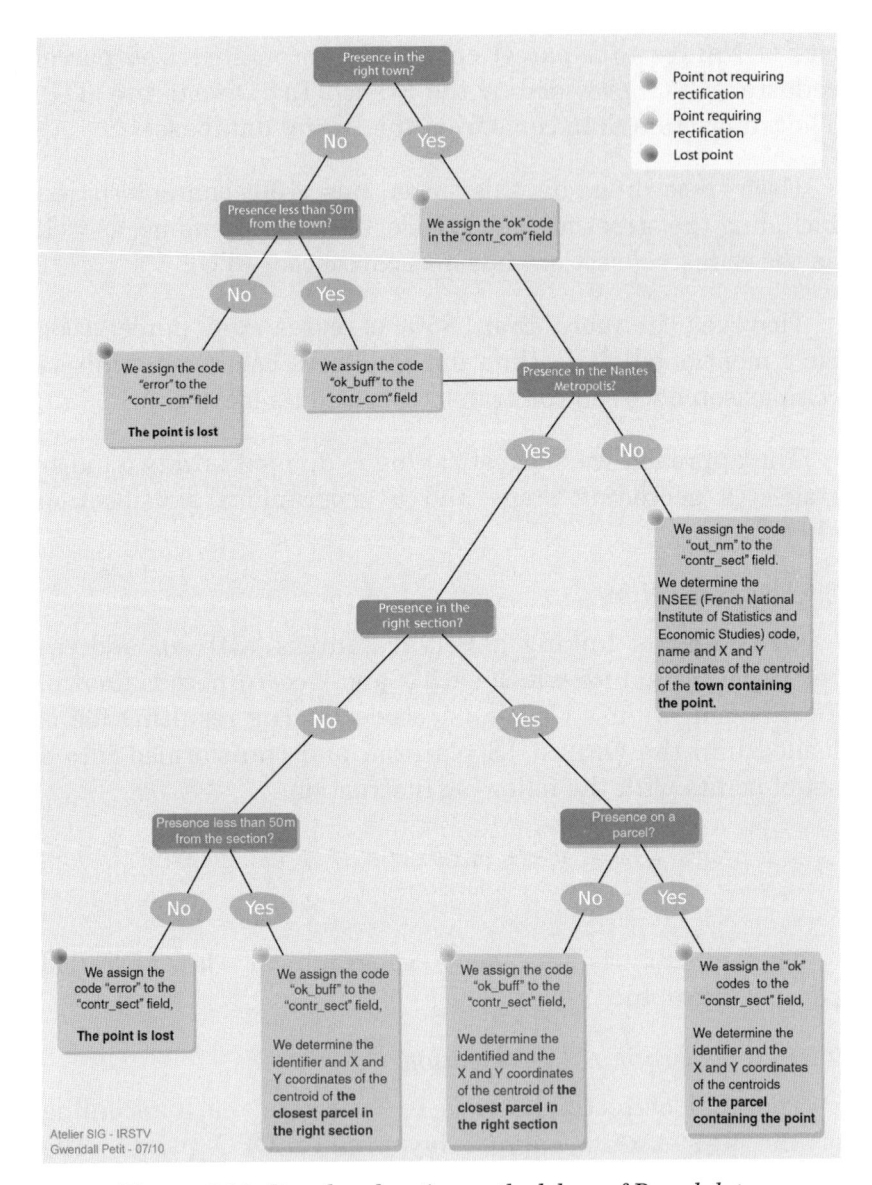

Figure 2.12. *Parcel geolocation methodology of Perval data*

For this sequence of processes, we use spatial operators and predicates, such as ST_Intersects, ST_Contains, and ST_Buffer. The imbrication of these operators allows us, for example, to determine the uncertainty distances within which points of the pervalGeographic layer can be rectified.

When a pervalGeographic point is less than 50 m from the town it belongs to, and the town's name is in the Perval database, then it is considered to be "rectifiable". If the point is beyond the uncertainty distance of 50 m, then the geocoding error is too big and the point will not be rectified. A contr_com field is added and assigned the error value. If the point is within the uncertainty distance, then the ok_buff value is assigned to the contr_com field. If the pervalGeographic point is located within the limits of the town it belongs to, then the value ok is added in the contr_com field.

The method then consists of iterating the processes by refining the spatial and semantic queries. We thus use the section number information to assert that the point has been located. In the same way, to look for the closest parcel, we will use a distance analysis function ST_NearestPoints (Figure 2.13).

We present below a selection of spatial SQL queries that are representative of the Perval data rectification chain.

```
-- Creating points to be rectified from two X and Y fields
-- contained in the "pervalGeocode" field.
CREATE TABLE pervalGeographic AS SELECT ST_MakePoint(X, Y)
   AS the_geom, * FROM pervalGeocode;

-- Creating the <<contr_com>> field and analyzing
    it versus
-- an uncertainty distance.
ALTER TABLE pervalGeographic ADD COLUMN contr_com TEXT;
UPDATE pervalGeographic SET contr_com = 'ok_buff'
```

```
    WHERE pervalGeographic.num_act_perv IN
    (SELECT a.num_act_perv
        FROM pervalGeographic AS a, communes_au as b
        WHERE ST_INTERSECTS(b.the_geom,
            ST_BUFFER(a.the_geom,50))
        AND a.num_cominsee=b."CODE_INSEE"
    );

-- Calculating the distance between each point in the
    Perval base and
-- the closest points of each surrounding parcel
-- The result is valid only if the point is in
-- the correct cadastral section (contr_sect='ok_buff')
CREATE TABLE dist_parc_ pervalGeographic AS
    SELECT ST_NEARESTPOINTS(a.the_geom, b.the_geom) AS
    geom_result, a.num_act, b.IDOBJ
    FROM pervalGeographic a, parc_in_buff_appart b
    WHERE a.contr_sect='ok_buff'
    AND a.cod_section=b.cod_section
    AND ST_INTERSECTS(ST_BUFFER(a.the_geom, 50),
        b.the_geom);
```

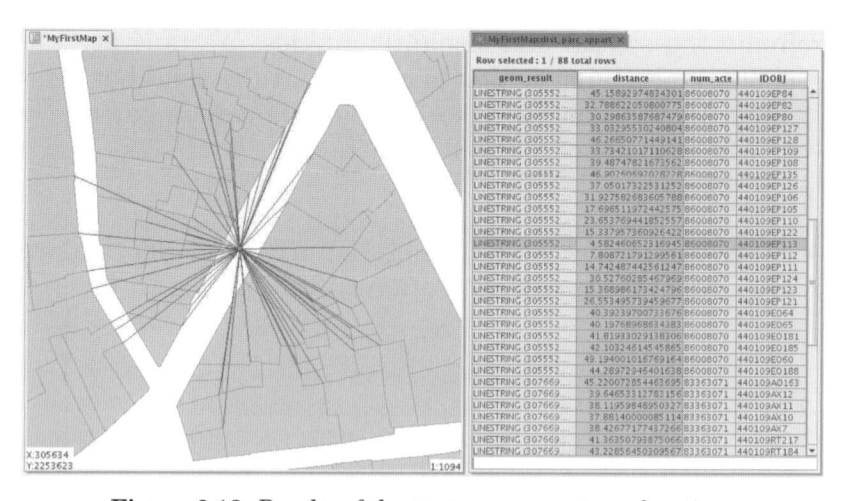

Figure 2.13. *Results of the* ST_NearestPoints *function*

Figure 2.14 represents the result of the Perval data rectification to the parcel with a classification of the three types of transaction specified in the base: apartment, house, and land. A percentage between the number of rectified points and original data is given as an indication.

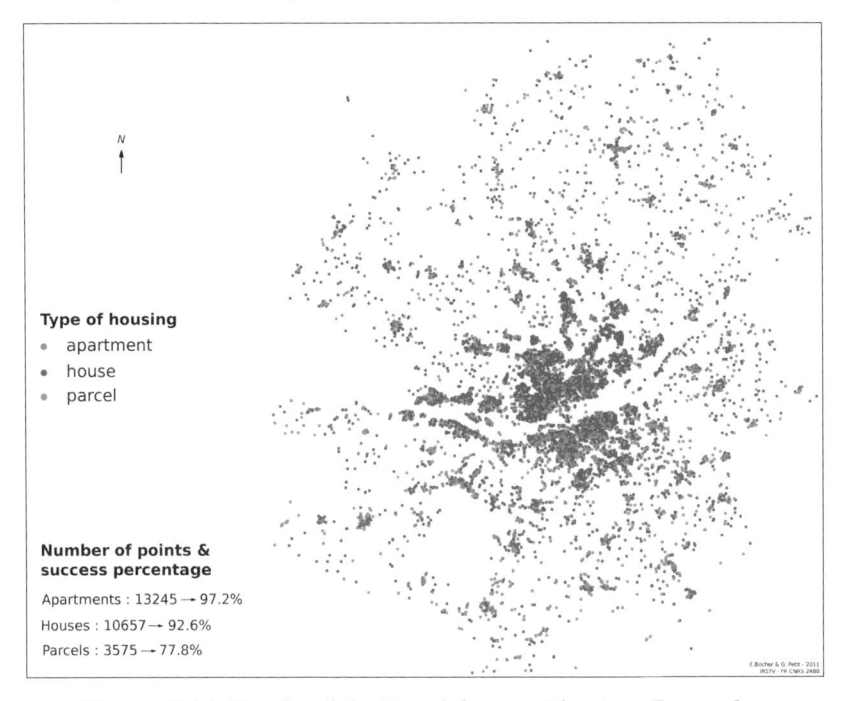

Figure 2.14. *Results of the Perval data rectification. For a color version of this figure, see www.iste.co.uk/Bucher/innovgis.zip*

2.6. Community

The ORBISGIS community is made up of developers and users that can be found on the project mailing list[13]. It currently has 80 subscribers.

The GIS platform that coordinates the ORBISGIS development also takes part in research programs to help the

---

13 http://www.orbisgis.org

platform evolve. Table 2.6 provides a list of research programs and topics for which ORBISGIS is used.

Program name	Application topic
ANR AVUPUR	Spatial hydrology
ANR EvalPDU	Noise map and indicator production
ANR EvalPDU	Spatial econometrics
ANR VEGDUD	Spatiotemporal model
SCAPC2	Cartography and interoperability
SOGVILLE	Cartography, language, and geoprocessing service
GEBD NETWORK	Retropolated repository and graph analysis

Table 2.6. *List of research programs using* ORBISGIS

Moreover, the GIS platform spurs collaborations with European laboratories and universities as well as local authorities and state services. The GIS platform has, for example, taken part in the organization of a summer school on free and *open source* GIS spearheaded by the SIGTE Laboratory of the University of Girona in Spain[14].

Let us conclude by mentioning the strong ties between the IRSTV and the geomatics actors in the Pays de la Loire region through the GÉOPAL network[15]. This network is a direct link between the research activities carried out around the ORBISGIS platform and their use by local authorities. GÉOPAL thus finances important developments to improve map production and exchange within the context of an SDI (SOGVILLE project, 2010–2013).

For further information, the reader can visit the www.orbigis.org website.

---

14 http://www.sigte.udg.edu/summerschool2010/, accessed September 2011.
15 http://www.geopal.org/accueil, accessed September 2011.

2.7. Conclusion and perspectives

The ORBISGIS platform offers all the main GIS functionalities: visualization, shifting, interrogation, and geometric and alphanumerical editing. Its SQL grammar allows us to design advanced processing and analysis chains. We have provided a few examples. Conceived and designed in a modular fashion, the ORBISGIS graphical interface can be extended with *plugins*. Thus, a noise map calculation module is currently being developed by the "Acoustics" team at the UR EASE – IM Department of IFSTTAR, within the framework of the French National Research Agency's EvalPDU. That said, in spite of these advantages, there is still a lot of work to do in order to answer interoperability requirements and representation and modeling needs that have emerged from research programs. Let us mention:

– reading new data fluxes: Web Feature Service (WFS), Table Join Service (TJS);

– taking geographic projections into account;

– improving the cartography interface, especially the *Symbology Encoding*[16] specification used to publish representations on a cartographic server;

– enriching the SQL language to interrogate new data structures: network graphs, triangulated irregular network (TIN), etc.;

– modeling temporal series and reasoning based on them.

These points are all part of ORBISGIS version 4.0's roadmap.

---

[16] http://www.opengeospatial.org/standards/symbol, accessed September 2011.

2.8. Acknowledgments

The authors gratefully thank the entire management team at the IRSTV, whose work ensured that the ORBISGIS platform would be developed and deployed in excellent material and human conditions. The authors also thank the IRSTV members who support the creation of this collaborative GIS through their research programs. Finally, the authors would like to thank all the free and *open source* community.

2.9. Bibliography

[BOC 07a] BOCHER E., Projet d'infrastructure de données spatiales, LCPC, Nantes, France, March 2007.

[BOC 07b] BOCHER E., LEDUC T., GONZALÉZ-CORTÉS F., "OrbisGIS: a GIS for scientific simulation", *Rencontres Mondiales du Logiciel Libre (RMLL)*, Amiens, France, 2007.

[BOC 07c] BOCHER E., LEDUC T., LONG N., GONZALÉZ-CORTÉS F., "UrbSAT: outil d'extraction d'informations géographiques et de production d'indicateurs", *Journées Information géographique et observation de la terre – GDR SIGMA-Cassini et I3*, Strasbourg, France, 2007.

[BOC 08a] BOCHER E., Audit sur les données du Secteur Atelier Pluridisciplinaire, Internal report, IRSTV, 2008.

[BOC 08b] BOCHER E., LEDUC T., MOREAU G., GONZALÉZ-CORTÉS F., "GDMS: an abstraction layer to enhance Spatial Data Infrastructures usability", *Proceedings of the 11th AGILE International Conference on Geographic Information Science*, Girona, Spain, 2008.

[CLI 94] CLINTON W., Coordinating geographic data acquisition and access to the National Spatial Data Infrastructure, Executive Order 12096, Federal Register 59, Washington, DC, USA, April 1994.

[EGE 88a] EGENHOFER M.J., FRANK A.U., "Towards a spatial query language: user interface considerations", *Proceedings of the 14th International Conference on Very Large Data Bases, VLDB '88*, San Francisco, CA, USA, pp. 124–133, 1988.

[EGE 88b] EGENHOFER M.J., FRANK A.U., "Designing object-oriented query languages for GIS: human interface aspects", *Third International Symposium on Spatial Data Handling*, Sydney, Australia, pp. 79–97, 1988.

[EGE 89] EGENHOFER M., Concepts of spatial objects in GIS user interfaces and query languages, Report no. 90-12, National Center for Geographic Information and Analysis, 1989.

[GOH 89] GOH P.-C., "A graphic query language for cartographic and land information systems", *International Journal of Geographical Information Science*, vol. 3, no. 3, pp. 245–255, 1989.

[GUT 88] GUTING R., "Geo-relational algebra: a model and query language for geometric database systems", *Computational Geometry and Its Applications*, LNCS 333, Springer, Berlin, Heidelberg, pp. 90–96, 1988.

[HÉG 06] HÉGRON G., MEIGEVille, Modélisation environnementale intégrée et gestion durable de la ville, Appel à projets régionaux, IRSTV & Région des Pays de la Loire, 2006.

[HER 99] HERRING J., JOHNSON S., OpenGIS Simple Features Implementation Specification for SQL, Open Geospatial Consortium, 1999.

[HER 06a] HERRING J., OpenGis implementation specification for geographic information – simple feature access – part 1: common architecture, Open Geospatial Consortium, 2006.

[HER 06b] HERRING J., OpenGis implementation specification for geographic information – Simple feature access – part 2: SQL option, Open Geospatial Consortium, 2006.

[INS 07] Parlement Européen & Conseil, Directive 2007/2/EC (INSPIRE) établissant une infrastructure pour l'information spatiale dans la Communauté européenne, 2007.

[LED 09] LEDUC T., BOCHER E., GONZALÉZ-CORTÉS F., MOREAU G., "GDMS-R: a mixed SQL to manage raster and vector data", *Proceedings of GIS 2009*, Ostrava, the Czech Republic, 2009.

[NEB 04] NEBERT D.D., (ed.), *Developing Spatial Data Infrastructures: The SDI Cookbook*, Technical Working Group Chair, GSDI, January 2004.

[OCA 84] OCALLAGHAN J., MARK D., "The extraction of drainage networks from digital elevation data", *Computer Vision, Graphics, and Image Processing*, vol. 28, no. 3, pp. 323–344, 1984.

[PLA 02] PLANCHON O., "A fast, simple and versatile algorithm to fill the depressions of digital elevation models", *CATENA*, vol. 46, nos. 2–3, pp. 159–176, 2002.

[TOM 79] TOMLIN C.D., BERRY J.K., "A mathematical structure for cartographic modeling in environmental analysis", *Proceedings of American Congress on Surveying Mapping*, Falls Church, VA, USA, pp. 269–283, 1979.

Chapter 3

GEOXYGENE: an Interoperable Platform for Geographical Application Development

3.1. Introduction

Geographical application development generates costs for research laboratories [BAD 03, BUC 11]. In spite of standardization efforts made by consortia such as ISO[1] and OGC[2], the different geographical models implemented in common GIS software do not allow for interoperable use of these models. The applications developed for one of these non-standard models are thus not necessarily reusable for applications based on a different model. Moreover, the programming languages used to develop applications relying on market GIS software are often "legacy" languages. Thus, sharing code becomes difficult and methods developed for one

---

Chapter written by Éric GROSSO, Julien PERRET and Mickaël BRASEBIN.
1 International Organization for Standardization: http://www.iso.org.
2 Open Geospatial Consortium: http://www.opengeospatial.org.

software have to be reimplemented in other software in order to be used. It is a time-consuming process for developers as well as for reseachers who have to learn new programming or GIS environments. These reasons were behind the decision of many laboratories to choose solutions that used free software. More specifically, the COGIT Laboratory of the IGN (French Mapping Agency) began in 2000 to develop the *open source* GEOXYGENE platform.

3.2. Background history

Originally designed between 2000 and 2004 by Thierry Badard and Arnaud Braun, then enriched by many researchers at the COGIT Laboratory, the GEOXYGENE platform's first release (1.0) was registered[3] in 2005 under an LGPL (GNU *Lesser General Public License*)[4]. Its main goal is to respond to and meet the laboratory's development needs. These needs are many and varied, and concern geographical data interoperability as much as the reuse of the data, the maintenance and sharing of code between researchers (laboratory and external researchers), as well as the manipulation of the used data model. Indeed, the question was both to improve the development of research applications and to better understand data and the models used to manipulate it.

The different GEOXYGENE releases are:

– Release 1.0: registering the GEOXYGENE core (May 2005);

– Release 1.1: correcting minor bugs (June 2006);

---

3 http://oxygene-project.sourceforge.net/
4 http://www.gnu.org/copyleft/lesser.html

– Release 1.2: registering "basic" applications (August 2007);

– Release 1.3: registering the data matching tools (January 2008) [MUS 08];

– Release 1.4: registering (June 2009):

- data persistence with Hibernate[5] (as well as OJB[6]);

- new graphical interface;

- management of complex styles for map display;

– GPL registering for the interface between GEOXYGENE and OpenJUMP (June 2009);

– Release 2.0 (throughout 2012):

- GPL registering of the three-dimensional (3D) model [BRA 09];

- legend design support and improvement [BUA 09, CHR 09];

- data matching based on the belief function theory [OLT 09].

3.3. Major functionalities and examples of use

GEOXYGENE's major functionalities concern the loading and manipulation of geographical data compatible with ISO norms and OGC standards. GEOXYGENE was designed as a research application platform. The targeted audience is thus essentially made up of the geomatics researcher community and, more generally, of GIS application developers. GEOXYGENE also allows us to develop more

---

5 http://www.hibernate.org/
6 http://db.apache.org/ojb/

collaboration possibilities between former doctoral candidates and laboratory researchers.

3.3.1. *Generic functionalities*

GEOXYGENE provides the user with tools that enable the loading and saving of data from/in an *ESRI Shapefile* format, as well as storing spatial data in a POSTGRESQL/POSTGIS database[7]. Moreover, GEOXYGENE offers algorithms allowing the user to create and manipulate topological maps, match databases, create and improve legends, etc., and allows the user, if the existing algorithms do not correspond to his/her needs, to develop new algorithms. Finally, it is possible to visualize and edit data and their schemas due to a dedicated graphical interface or an *ad hoc plugin* for OpenJUMP[8], another *open source* GIS software.

3.3.2. *Use case: building data manipulation*

The use case is described due to two fictional characters, called Justine and Robert: "Justine, an experienced developer, has data in the *ESRI Shapefile* format which describes the buildings in a given geographical zone (in this case, Orléans) and wishes to load and manipulate it, run her algorithms, and add the results of these algorithms to her data schema to then send it to Robert, who does not do much developing".

3.3.2.1. *Data*

In this example, our starting hypothesis is that the user (Justine) has the data in *ESRI Shapefile* format files. The data illustrated in Figures 3.1 and 3.2 correspond to the BDTOPO® database that contains the description of buildings respecting

---

7 http://www.postgis.fr
8 http://www.openjump.org

precise specifications; they must, for example, have a surface superior to 20 m². In our example, we have used Release 1.3 of these database specifications.

Figure 3.1. *Attribute data on the* BDTOPO® *buildings*

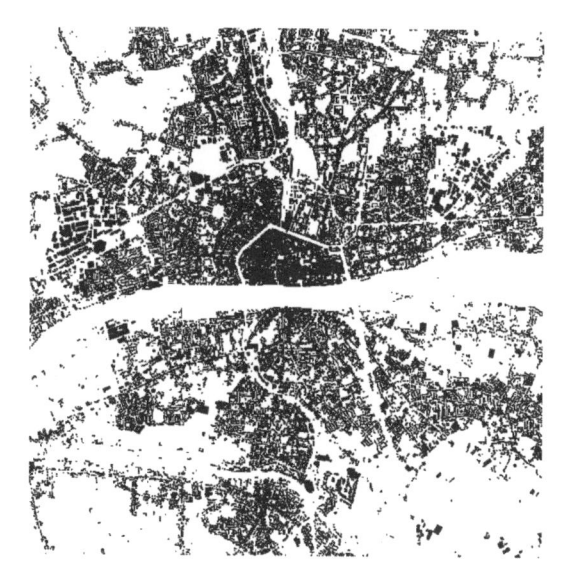

Figure 3.2. *Geometrical data on the* BDTOPO® *buildings*

First of all, the user must load her data into a database management system. For example, Justine uses a POSTGRESQL/POSTGIS database called geoxygene. This load can, for example, be carried out in a command line with the `shp2pgsql` function. The complete command will look like this:

```
shp2pgsql -g geom -D -I building.shp buildings
           | psql geoxygene
```

It is a double command. The part before "|" corresponds to the creation of an SQL loading command file from the file in the *ESRI Shapefile* format. The parameters mean that the geometry is called `geom`, that the loading format to be used is `dump`, and that an index on the geometry column must be created, and that the new table will be called `buildings`. The part after "|" corresponds to the execution of the SQL commands written in the generated file to create a table in the `geoxygene` base.

Moreover, GEOXYGENE is the user to directly load data in an *ESRI Shapefile* format on volatile memory, without having to go through intermediate storing in a database (as mass storage). However, in this example, the user would rather use a database to take advantage of its functionalities (concurrent access, transaction security, "random access" to avoid an "out of memory" error when loading an amount of data exceeding the workstation's volatile random access memory capacity, etc.).

3.3.2.2. *The data schema: the* `Building` *class*

Using GEOXYGENE tools, Justine generates a fitting JAVA class from the loaded database, illustrated in Figure 3.3, as well as a *mapping* file that will match the JAVA class attributes with the columns of the POSTGRESQL/POSTGIS database tables (see Figure 3.4). The data schema created fits with the initial data. Justine can then modify this schema to add relative attributes to the processing results.

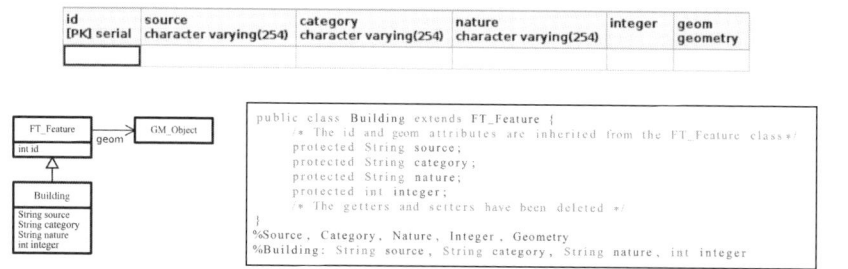

id	source	category	nature	integer	geom
[PK] serial	character varying(254)	character varying(254)	character varying(254)		geometry

```
public class Building extends FT_Feature {
    /* The id and geom attributes are inherited from the FT_Feature class */
    protected String source;
    protected String category;
    protected String nature;
    protected int integer;
    /* The getters and setters have been deleted */
}
%Source, Category, Nature, Integer, Geometry
%Building: String source, String category, String nature, int integer
```

Figure 3.3. *Automatic generation of the JAVA structure fitting with the building objects originally stored in a POSTGRESQL/POSTGIS database. For a color version of this figure, see www.iste.co.uk/Bucher/innovgis.zip*

```
<class-descriptor class="data.Building" table="building" >
  <field-descriptor name="id" column="ID" jdbc-type="INTEGER" primarykey="true" autoincrement="true"/>
  <field-descriptor name="source" column="source" jdbc-type="VARCHAR" />
  <field-descriptor name="category" column="category" jdbc-type="VARCHAR" />
  <field-descriptor name="nature" column="nature" jdbc-type="VARCHAR" />
  <field-descriptor name="integer" column="integer" jdbc-type="INTEGER" />
  <field-descriptor name="geom" column="geom" jdbc-type="STRUCT"
                    conversion="fr.ign.cogit.geoxygene.datatools.ojb.GeomGeOxygene2Dbms" />
</class-descriptor>
```

Figure 3.4. *Example of object-relational mapping file with OJB. For a color version of this figure, see www.iste.co.uk/Bucher/innovgis.zip*

3.3.2.3. *Object-relational mapping with OJB*

Justine can then implement her processing algorithms inside the `Building` class using its attributes (and thus the data model) while using the persistence, i.e. saving the results coming from the processing in the `geoxygene` database. And to provide the manipulated data to Robert, Justine can, for example, choose to save the data in an *ESRI Shapefile* format, or provide him with access to her POSTGRESQL/POSTGIS database through the Internet network (using Web services, for example).

3.3.2.4. *A processing example: building urban areas*

A simple and common process (illustrated in Chapter 2 on ORBISGIS software) is to create urban areas from buildings. Such a process means merging the buffer zones of a certain size (in our example, 50 m), created from the geometry of buildings. Once these zones are merged, they are called urban areas and can be subject to more processes (Douglas–Peucker filtering to simplify geometry, for example), linked to the buildings they belong to, and finally qualified (by area, number of buildings they contain, etc.) [BOF 01]. The created data layer is illustrated in Figure 3.5. GEOXYGENE allows for the storing of this new layer in a database table or in a file under the *ESRI Shapefile* format.

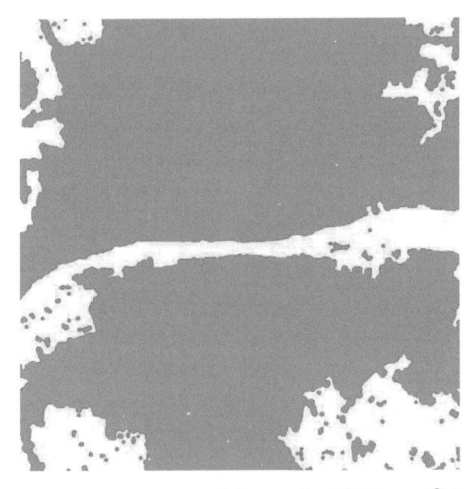

Figure 3.5. *Urban areas created from the* BDTOPO® *buildings (see Figures 3.1 and 3.2)*

One of the most common tools used in GEOXYGENE is the topological map. Mainly developed by Sébastien Mustière and Olivier Bonin, it offers an alternative to the topological model defined by the ISO 19107 norm, considered too cumbersome. The topological map actually refers to a topological structure, allowing us to import objects (point, line, or surface) and treat them as topological objects (vertices, edges, or faces). Its main advantage is that it enables us to use a set of methods to create missing vertices, merge the duplicate vertices, delete the simple nodes (vertices with degree 2), and compute the associated planar graph (cutting up the edges, correcting the geometry of the objects, etc.). A processing example (illustrated in Figure 3.6), and we need to import the roads, paths, railways, and waterways of the BDTOPO®, and create a planar graph used here to divide the urban areas into urban blocks.

Figure 3.6. *Urban areas divided into urban blocks due to the linear objects of* BDTOPO® *and the topological map*

3.4. Architecture

In GEOXYGENE, the geographical data manipulation is carried out through an application schema adapted to user's needs. The link between this application schema and the original data schema (coming from a data producer or another user) is made possible due to an object-relational mapping. To carry out this match within GEOXYGENE, common tools such as OJB[9] and Hibernate[10] are used. This mapping is explained in section 3.3.2.3.

---

9 http://db.apache.org/ojb
10 http://www.hibernate.org

76 Innovative Software Development in GIS

The three applicative layers for the development of applications in GEOXYGENE, illustrated in Figure 3.7, are the following:

– the GEOXYGENE core, containing main data structures;
– the basic data manipulation applications;
– the expert applications resulting from research work.

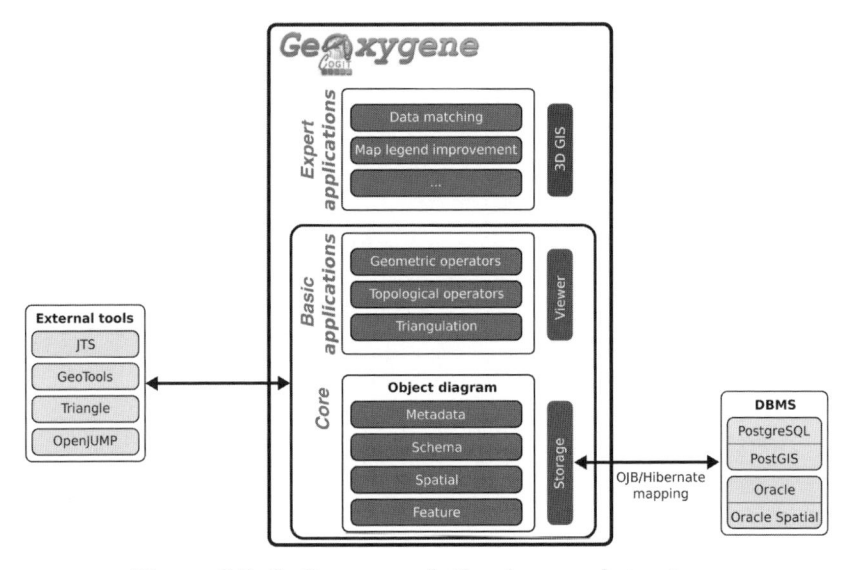

Figure 3.7. GEOXYGENE *platform's general structure*

3.4.1. *The core*

The main functionalities of GEOXYGENE's core concern the modeling of the applicative object schema. This layer enables the representation of data, the elements related to it, such as its geometry, its topology, its attributes, and its metadata, as well as operations directly applicable to data modeled by spatial objects (*features*). The core also contains tools enabling the import or export of these data toward databases as well as a certain number of external tools (including GEOTOOLS and OpenJUMP).

Moreover, to guarantee the interoperability with norms and standards related to geographical information, the core implements them partially:

ISO 19107: geometry and topology representation. This is a set of *packages* (groups of classes) gathered in the "Spatial" module. The norm has been slightly modified to allow for greater ease of use [ISO 03a].

ISO 19109: metamodel allowing us to build a geographical schema, also known as metamodel for geographical classes. This is a set of metaclasses gathered in the "Dico" *package*. These classes are the data dictionary (the types of attributes, all the possible values, etc.) [ISO 05].

ISO 19115: metadata. These classes are gathered in the "Metadata" *package* [ISO 03b]. Only the most relevant classes have been implemented, those that match what can generally be found at the IGN or outside the Institute. The evolution of this implementation will depend on the research carried out at the COGIT Laboratory on consulting and unifying geographical databases.

3.4.2. *First applicative layer: the basic applications*

This applicative layer contains many data manipulation tools:

– geometric operators (angles, vectors, and algorithmic geometry) due to the *JTS Topology Suite* library [11];

– topological operators (graph algorithms);

---

11 http://sourceforge.net/projects/jts-topo-suite/

78 Innovative Software Development in GIS

– Delaunay triangulation due to the TRIANGLE library [SHE 96][12].

This applicative layer also offers a simple graphical interface (see Figure 3.8), which completes the series of tools enabling users to visualize and explore their data.

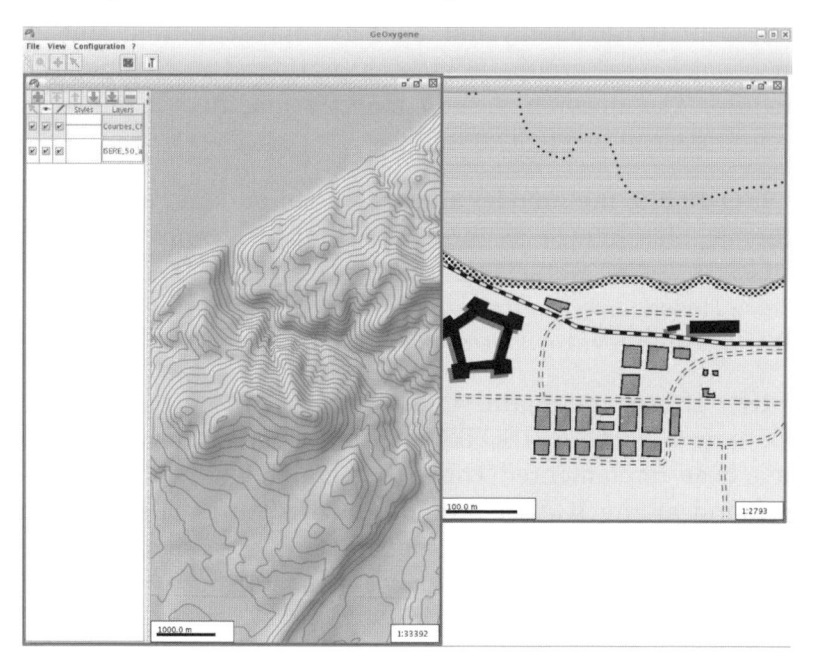

Figure 3.8. *Graphical interface of the* GEOXYGENE *platform*

3.4.3. *Second applicative layer: the expert applications*

This layer contains the research applications licensed under *open source*. It notably contains the data matching

---

12 Let us note that this library is not registered under an *open source* license. GEOXYGENE offers the possibility of using TRIANGLE due to a dynamic link, but the user has to obtain it himself: http://www.cs.cmu.edu/quake/triangle.html.

algorithms for multilevel data, such as networks [DEV 97, MUS 08] and surfaces [BEL 01]. Tools are also provided to allow for storing, management, and visualization of the data matching links that are created.

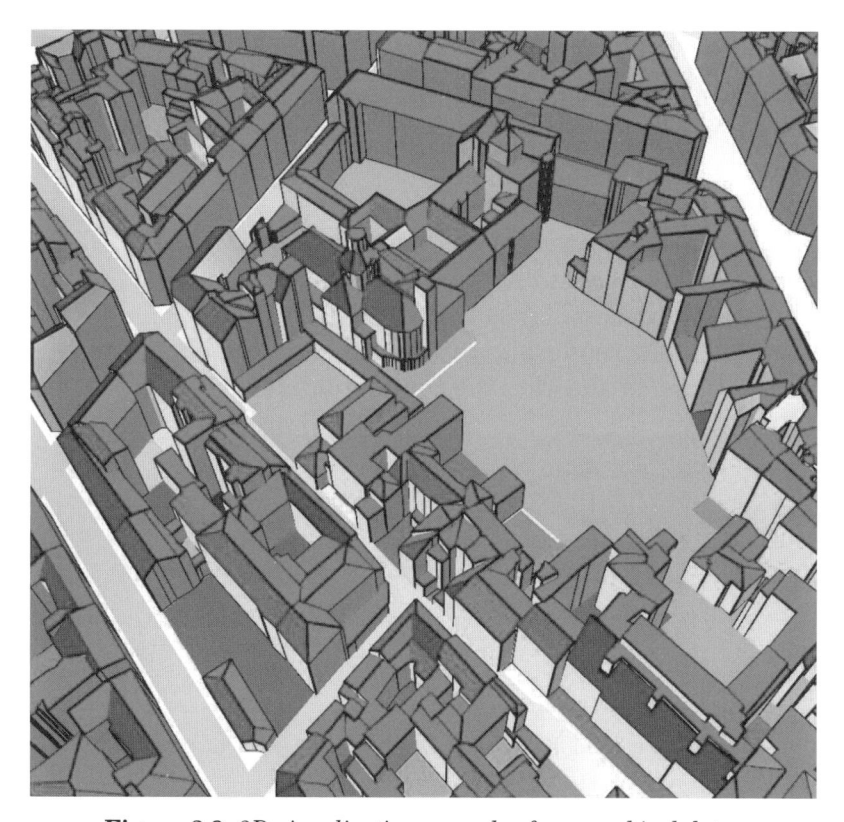

Figure 3.9. *3D visualization example of geographical data*

Among the applications currently being developed, and which will be licensed soon, there are tools that

– visualize and manipulate 3D data [BRA 09] (see Figure 3.9);

– manipulate geographical and ontological schemas [ABA 09];

– edit metadata: create a set and its metadata;

– analyze hydrographic graphs to characterize the landscape [PAG 08];

– improve the colors used to represent data [BUA 09, CHE 06];

– use a system of map specifications and legend improvements [CHR 09].

3.4.3.1. *Semiology modules*

The last two examples illustrated in Figure 3.10 are particularly important because they allow us to both question the design process for a map legend, provide simple and intuitive tools rendering this process accessible to general public, and also question the content of a legend.

Figure 3.10. *Map created from a Van Gogh painting, according to [CHR 09]*

Moreover, due to the implementation, manipulation, and extension of different standards linked to map symbology (in this case OCG Filter [FIL 05] and Symbology Encoding [OGC 05]), the laboratory is developing expertise in these standards and has become the source of proposals.

3.4.3.2. GEOXYGENE *3D module*

To carry out research on 3D, the laboratory decided to expand GEOXYGENE in order to integrate the third dimension. This expansion was developed to respect the choices made at the core level, and actually enabled the completion of geometrical classes to be in line with the ISO 19107 norm (taking the Z coordinates into account, as well as classes allowing us to model solids, etc.).

This geometrical model can be instantiated due to different loaders integrated by the module. Different formats are also available, either from the 2D data or from 3D data that are transformed (e.g. by extrusion): CityGML[13], digital terrain models – or DTM – (file in a ARC/INFO ASCII GRID format), *ESRI Shapefile* that has been extruded or mapped on a DTM (see Figure 3.11), POSTGRESQL/POSTGIS data, or data coming from 3D modeling software (3DS, OBJ).

Beyond visualization and 3D data loading, the module also offers a series of 3D geometrical analysis functions, such as Boolean operators (addition, subtraction, intersection, union, etc.), intervisibility calculation (see Figure 3.12), computation of 3D buffers, volumes, and surface areas, or even the breaking down of volumes into triangles and tetrahedra.

The 3D module can be used as an API or an independent application. Indeed, it has a graphical interface allowing us to carry out the operations we have listed as well as the standard GIS operations of moving, representing, and consulting. The JAVA 3D[14] graphical library enables us to manage the 3D display. The module has already been used for various research works concerning the simplification of buildings, the suggestion of a management tool for urban planning rules [BRA 11], and the proposal for a visual variable enabling us to synthesize information carried by points of interest [BRA 10].

---

13 http://www.citygml.org
14 http://java3d.dev.java.net

82   Innovative Software Development in GIS

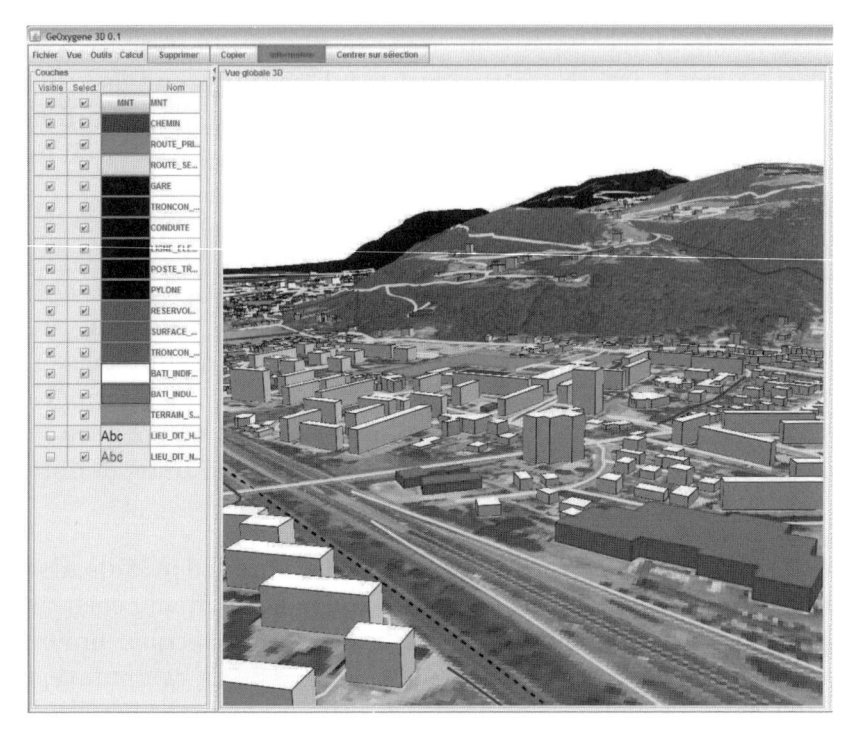

Figure 3.11. *Example of a 3D representation of the French "Large Scale Repository" (RGE). For a color version of this figure, see www.iste.co.uk/Bucher/innovgis.zip*

In the future, this module should welcome new 3D research, and specifically those linked to spatial analysis, as well as 3D semiology including, for example, geographical name placing, as well as new 3D-specific non-photorealistic (or expressive) styles proposals (see Figure 3.13).

3.4.3.3. GEOXYGENE *spatiotemporal module*

Another expert application developed on GEOXYGENE is the GEOPENSIM[15] platform. It has been developed within the framework of the eponymous project funded by the

---

15 http://geopensim.ign.fr

French National Research Agency (ANR), and aims to offer an analysis and simulation tool for the evolution of urban fabric. Within this framework, many new functionalities were developed, especially in the area of creation and representation of historical databases [PER 09], their analysis, and their simulation [PER 10] (see Figure 3.14). Let us point out that, in this figure, the densification of the urban blocks is exaggerated in order to make the process more legible.

Figure 3.12. *3D intervisibility calculation: the gray buildings are visible from the fountain in the middle of the square, depending on the view angle represented here by a prism*

Figure 3.13. *Representation of a building with stylized edges*

3.5. Communities

Since GEOXYGENE is used as a research platform in most of the work carried out at the COGIT, the main GEOXYGENE user and developer community is made up of the researchers from this laboratory. This community originally developed the core and basic applications for GEOXYGENE (see sections 3.4.1 and 3.4.2), and designed and implemented expert applications such as a generic multicriteria matching process [OLT 08], a tool to understand and interpret representation differences between pieces of geographical data [SHE 09] and the expert applications mentioned in the previous section (see section 3.4.3).

This community has thus opened itself up when doctoral candidates left for other laboratories, since they were able to continue using the applications they had developed. Indeed, the first and foremost goal of GEOXYGENE as a community is to create relationship between former doctoral candidates and laboratory researchers.

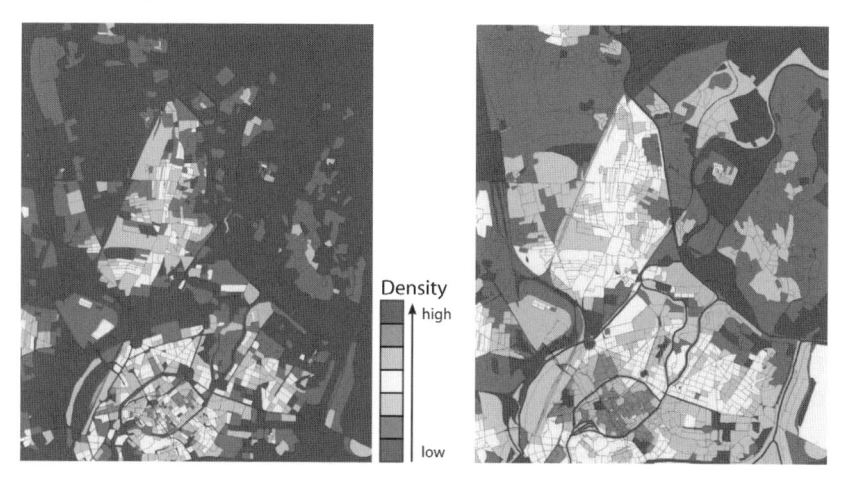

Figure 3.14. *Evolution rule application in the* GEOPENSIM *city blocks between 2002 and a simulated date in 2010. For a color version of this figure, see www.iste.co.uk/Bucher/innovgis.zip*

The GEOXYGENE user community then got noticeably richer, due to the expert applications being available (such as the network matching tool). This tool first of all was integrated internally within the IGN through the "New Topographic Map" project whose main goal is to implement a new production chain for the Basic Map (scale 1:25000) from existing IGN databases. The data matching tool is used to ensure geometrical consistency between different databases. The Belgian IGN then integrated the matching tool in its own production lines, to match its data at the scale of 1:10000 and 1:50000 to ease their update. Finally, the English Ordnance Survey and Wallonia (Belgium) are currently testing the tool in order to use it in the future [REV 09]. These first uses involve tasks such as advising, supporting, and even bug correcting. They allow us to highlight the limits of the proposed tools in terms of ergonomics or strength, and to improve them. These are actually excellent opportunities to create relationships between laboratories which might lead to fruitful scientific collaborations.

Beyond the specific use of GEOXYGENE as has been previously mentioned, the platform is used in its entirety in many projects, such as the GIS laboratory of the Ecole Polytechnique Fédérale de Lausanne, which uses GEOXYGENE, for its urban fabric simulations, and the Laval University in Canada, which uses it for teaching purposes.

The developer community has lately grown under the impetus of ANR GEOPENSIM project which has an aim to develop spatial and simulation analysis tools on GEOXYGENE.

Finally, the user community grows through GEOXYGENE training sessions carried out within the framework of the MAGIS (formerly SIGMA)[16] research group.

3.6. Conclusion

There are many advantages in having a GIS development platform in a research laboratory. The GEOXYGENE platform enables COGIT to spread its research results. The first level of dissemination obviously concerns the laboratory researchers. This dissemination is important because it allows researchers to capitalize and share developed tools on an internal level, and avoid multiple implementations of the same process. This dissemination also allows for the value of the tools developed during doctoral studies to increase, and for these tools to

---

16 http://magis.ecole-navale.fr

be reused. A second level of dissemination concerns all the laboratory partners, within research projects, for example, as well as other IGN departments, especially production, for which the transfer of tools developed in GEOXYGENE (such as data matching) is planned out.

One of the strongest feedbacks about GEOXYGENE concerns the importance of the graphical interface for users, at all levels of development expertise. Indeed, until recently, GEOXYGENE only had a basic interface to visualize geographical data. For example, this interface did not offer users any simple solution to run their algorithms. To overcome this deficiency, we developed an interface between GEOXYGENE and OpenJUMP; however, the use of different data structures, although it enabled us to demonstrate the interoperability of both tools, has not allowed for an optimal use of the resources. As such, the laboratory is currently developing a graphical interface for GEOXYGENE. This interface, due to the implementation of OGC Filter Encoding and Symbology Encoding standards, allows us to describe complex styles for map display. It also allows users to run simple processes and has the particularity of being easily modular and extensible, as the GEOPENSIM simulator development proves.

Future developments in GEOXYGENE will focus on users, especially due to the new interface and the release of urban fabric analysis tools developed in the laboratory. Moreover, the tools to help design and improve map legends should allow users to create better maps without having any prior cartographic knowledge, and still allow them to express themselves creatively. Finally, the GEOPENSIM module, released in 2011, offers an *open source* tool to create historical databases, analyze the evolution of such databases, and simulate urban evolutions.

3.7. Bibliography

[ABA 09] ABADIE N., "Schema matching based on attribute values and background ontology", *12th AGILE International Conference on Geographic Information Science*, Hannover, Germany, 2009.

[BAD 03] BADARD T., BRAUN A., "Oxygene: une plate-forme interopérable pour le déploiement de services Web géographiques", *Bulletin d'Information Scientifique et Technique de l'IGN*, no. 74, pp. 113–122, 2003.

[BEL 01] BEL HADJ ALI A., Qualité géométrique des entités géographiques surfaciques. Application à l'appariement et définition d'une typologie des écarts géométriques, PhD Thesis, University of Marne-la-Vallee, 2001.

[BOF 01] BOFFET A., Méthode de création d'informations multiniveaux pour la généralisation de l'urbain, PhD Thesis, University of Marne-la-Vallee, 2001.

[BRA 09] BRASEBIN M., "GeOxygene: an open 3D framework for the development of geographic applications", *12th AGILE International Conference on Geographic Information Science*, Hannover, Germany, 2009.

[BRA 10] BRASEBIN M., BUCHER B., HOARAU C., "Enriching a 3D world with synthetic and visible information about the distribution of points of interest", *3D GeoInfo Conference*, Berlin, Germany, 2010.

[BRA 11] BRASEBIN M., PERRET J., HAËCK C., "Towards a 3D geographic information system for the exploration of urban rules: application to the French local urban planning schemes", *28th Urban Data Management Symposium (UDMS 2011)*, September 2011.

[BUA 09] BUARD E., RUAS A., "Processes for improving the colours of topographic maps in the context of map-on-demand", *International Cartography Conference (ICC'09)*, Santiago, Chile, 2009.

[BUC 11] BUCHER B., BRASEBIN M., BUARD E., GROSSO E., MUSTIÈRE S., PERRET J., "GeOxygene: built on top of the expertness of the French NMA to host and share advanced GI Science research results", *Geospatial Free and Open Source Software in the 21st Century*, LNG&C, Springer-Verlag, 2011.

[CHE 06] CHESNEAU E., Méthodologie d'amélioration des cartes de risque par analyse locale des contrastes cartographiques, PhD Thesis, University of Marne-la-Vallee, 2006.

[CHR 09] CHRISTOPHE S., "Making legends by means of painters' palettes", *Cartography and Art*, LNG&C, Springer, Berlin, Heidelberg, pp. 1–11, 2009.

[DEV 97] DEVOGELE T., Processus d'intégration et d'appariement de Bases de Données Géographiques; Application à une base de données routières multi-échelles, PhD Thesis, University of Versailles - Saint Quentin en Yvelines, 1997.

[FIL 05] OPEN GEOSPATIAL CONSORTIUM, OpenGIS filter encoding implementation specification, 1.1.0, OGC 04-095, 2005.

[ISO 03a] INTERNATIONAL ORGANIZATION FOR STANDARDIZATION (TC 211), ISO 19107:2003 Geographic information – Spatial schema, 2003.

[ISO 03b] INTERNATIONAL ORGANIZATION FOR STANDARDIZATION (TC 211), ISO 19115:2003 Geographic information – Metadata, 2003.

[ISO 05] INTERNATIONAL ORGANIZATION FOR STANDARDIZATION (TC 211), ISO 19109:2005(E) Geographic information – Rules for application schema, 2005.

[MUS 08] MUSTIÈRE S., DEVOGELE T., "Matching networks with different levels of detail", *GeoInformatica*, vol. 12, no. 4, pp. 435–453, 2008.

[OGC 05] OPEN GEOSPATIAL CONSORTIUM, OpenGIS symbology encoding implementation specification, 1.1.0, OGC 05-077r4, 2005.

[OLT 08] OLTEANU A.-M., MUSTIÈRE S., "Data matching – a metter of belief", *The International Symposium on Spatial Data Handling (SDH'2008)*, Montpellier, France, 2008.

[OLT 09] OLTEANU A.-M., MUSTIÈRE S., RUAS A., "Fusion des connaissances pour apparier des données géographiques", *Revue Internationale de Géomatique*, vol. 19, no. 3, pp. 321–349, 2009.

[PAG 08] PAGET A., PERRET J., GLEYZE J.-F., "La géomatique au service de la caractérisation automatique des réseaux hydrographiques", *Physio-géo, Géographie Physique et Environnement*, vol. 2, pp. 147–160, 2008.

[PER 09] PERRET J., BOFFET MAS A., RUAS A., "Understanding urban dynamics: the use of vector topographic databases and the creation of spatio-temporal databases", *24th International Cartography Conference (ICC'09)*, Mahdia, Tunisia, 2009.

[PER 10] PERRET J., CURIE F., GAFFURI J., RUAS A., "Un système multi-agents pour la simulation des dynamiques urbaines", *Les 18e Journées Francophones sur les Systèmes Multi-Agents (JFSMA'10)*, Santiago, Chile, 2010.

[REV 09] REVELL P., ANTOINE B., "Automated matching of building features of differing levels of details: a case study", *24th International Cartography Conference (ICC'09)*, Santiago, Chile, 2009.

[SHE 96] SHEWCHUK J.R., "Triangle: engineering a 2D quality mesh generator and delaunay triangulator", in LIN M.C., MANOCHA D. (eds), *Applied Computational Geometry: Towards Geometric Engineering*, LNCS 1148, Springer-Verlag, Berlin, Germany, pp. 203–222, 1996.

[SHE 09] SHEEREN D., MUSTIÈRE S., ZUCKER J.-D., "A data-mining approach for assessing consistency between multiple representations in spatial databases", *International Journal of Geographical Information Science*, vol. 23, no. 8, pp. 961–992, 2009.

Chapter 4

Spatiotemporal Knowledge Representation in AROM-ST

4.1. Introduction

Arising from frame languages [MIN 75], close to description logics [BAA 03] and ontological languages [MCG 04], object-oriented knowledge representation systems (also known as OOKRS) [DUC 98, NAP 04] describe, organize, and store knowledge by relying on the general principles of the *object* paradigm (notions of class, instance, specialization hierarchy). They have various inference mechanisms (inheritance, procedural attachment, filtering, classification) that allow them to clarify the knowledge and fill in any missing pieces.

Among the various OOKRS, AROM (an acronym for "Allier Relations et Objets pour Modéliser" (Ally Relations and Objects to Model) [PAG 00]) is distinctive due to its two

---

Chapter written by Bogdan MOISUC, Alina MIRON, Marlène VILLANOVA-OLIVIER and Jérôme GENSEL.

core complementary representation structures – classes and associations, like the entities used in UML class diagrams. Another particularity of AROM is the presence of an algebraic modeling language (AML). AMLs (AMPL[1], LINGO[2]) offer to represent equation, constraint, or query systems due to a formalism close to notations generally used in mathematics.

In AROM, the AML allows us to build algebraic expressions integrating the base types, managed by the system, and their associated operators, which considerably increases the system's expressiveness, defined as its description power in knowledge expression.

Indeed, these algebraic expressions allow us not only to formulate queries on an AROM knowledge base, but also to introduce into the model the value of an attribute at the definition creation stage, as well as to introduce constraints involving one or more attributes.

Moreover, since the AROM's type module is expandable, we can enrich the AML with new types and new operators. This enabled us to expand AROM and add spatiotemporal operators, and use it as a modeling and data management tool for applications with spatiotemporal data. AROM-ST thus allowed us to fill certain expressiveness shortcomings of the classical geographical information system (GIS) and database management system (DBMS); it is also an interesting alternative to conceptual modeling tools, such as MADS [SPA 07] and PERCEPTORY [BÉD 04], whose expressiveness is limited by an approach based on pictograms, and whose usability is also limited by code generation approach. In fact, AROM-ST has proved itself, and is used as a modeling and

---

1 http://www.ampl.com/
2 http://www.lindo.com/

data management tool for the spatiotemporal information system generator GENGHIS [MOI 08] (see Chapter 5).

This chapter presents AROM-ST, the spatiotemporal extension to AROM. AROM-ST is a spatiotemporal knowledge representation tool dedicated to modeling and data management for spatiotemporal applications.

The chapter is divided up as follows: section 4.2 presents AROM-ST's qualities that led us to suggest a time extension. Section 4.3 gives details of the characteristics of the AROM-ST extension. Section 4.4 showcases two AROM-ST extensions designed for the semantic Web: AROM-OWL and ONTOAST. Section 4.5 provides an insight into the software architecture behind the AROM platform. Section 4.6 describes the AROM-ST user community. Finally, section 4.7 concludes the chapter, and sketches a few future development directions for AROM-ST.

4.2. From AROM to AROM-ST

4.2.1. AROM *in context: a knowledge representation tool*

Among the different knowledge representation paradigms, object-oriented knowledge representation (OOKR) allows us to model knowledge relative to a specific application domain by relying on representation entities called "objects". Among these objects, we usually separate "classes" and "instances" (except for "prototype" languages for which these entities are merged). A class translates a concept, a family of individuals, providing a set of properties or "attributes" that characterize them. An instance represents an individual. An attribute is described by a name and a set of "facets".

The OOKRS are distinguishable by the facets they offer. The "typed facets" are essential because they specify the type and set (called "domain") of the attribute's possible values, or in the case of a multivalued attribute, the number of possible values. An attribute's type can be defined among those (*integer, boolean, string*, etc.) present in the programming language used for implementation. It can be elaborated from builders (such as *set-of* and *list-of*). It can also refer to a class present in the knowledge base. In that case, the attribute models the oriented relation of an ordinary binary relation established between the attribute class and the domain class designated by the type. "Inference facets" describe a way to obtain an attribute's value when the latter is not present. Default value, procedural attachment (which calls up a method), filtering (the attribute's value is the result of a query), and value passing (which assigns an attribute another attribute's value) are the major inference mechanisms designated by OOKRS facets. The "reflex facets" allow us to implement an action before or after accessing, modifying, adding, or deleting a value. Finally, secondary facets enable, for example, the association of an HTML description to an attribute.

Within a knowledge base, classes are organized in a specialization hierarchy that is shaped either as a tree or as a graph. Specialization is a partial order relation where each description of a class attribute is of the following: (1) the same as the description of this same attribute in the superclass, (2) has a type which is a subtype of the description of this same attribute in the superclass, and (3) corresponds to the definition of a new attribute. Indeed, class instances are instances of its superclass, specialization is equal to set inclusion. An inheritance mechanism adds itself to the specialization hierarchies, allowing us to factorize this information. These hierarchies are also the support for a core inference mechanism in OOKR: classification. It is of two types:

– "Instance classification", which aims to "bring down" an instance from its attachment class to its more specialized subclasses, by activating, if needed, the inference mechanisms required to determine missing attribute values.

– "Class classification", which determines the insertion position(s) of a class in the specialization hierarchy by looking for the more specialized superclasses and the more general subclasses.

Beyond type verifications related to any assignment of a value to an attribute, some OOKRS integrate consistency maintenance mechanisms that aim to disseminate the consequences of any change to an object of the knowledge base to adjacent objects.

4.2.2. *Originalities*

AROM [PAG 00] is a knowledge representation system in line with other OOKRS. It takes up their main principles: distinction between classes and instances, class specialization in an inheritance hierarchy, and presence of typed and inference facets. However, it is distinguishable from its predecessors in its way of representing relations between objects: AROM forbids the use of relational attributes (attributes typed by a class) and forces the modeling of relations between objects in an explicit manner, due to a second entity: association. An AROM-ST association is a named entity representing a set of n-tuple objects ($n \geq 2$) which it links. It is described by its roles (connection between the association and one of the linked classes) and by a set of variables (identical to UML attributes) defining properties related to the n-tuples. An inheritance of the roles, variables, and faces is possible between associations due to a specialization hierarchy similar to the class specialization hierarchy.

AROM's AML aims to define and manipulate the algebraic expressions contained in constants, variables, and operators of any type managed by AROM. By integrating the identifiers for classes, associations, objects (class instances), and tuples (association instances), these expressions allow us to question and use the entities defined in a knowledge base.

4.2.3. *Why a spatiotemporal extension?*

4.2.3.1. *Existence*

One of the solutions aiming to avoid the inconveniences of DBMS related to the integration of time and space comes from the conceptual modeling field [BÉD 04, SPA 07]. The emergence of conceptual modeling tools offers an interesting possibility to spatiotemporal data application designers. Indeed, the joint use of conceptual modeling tools with DBMS can ease designer's work in such applications and reduce the development time of these applications. The use of modeling tools also allows us to make up for certain shortcomings in terms of expressiveness at the DBMS level. For example, the MADS tool [SPA 07] allows us to express spatiotemporal constraints on object relations that are transformed by a code generator into integrity constraints at the level of the target DBMS.

MADS has revealed itself to be a very exhaustive and homogeneous representation model. The topological and time relation management through associations brings something extra to the table compared to other approaches, and associating generation with transition means we can model certain dynamic aspects of reality. The spatial and temporal pictograms are explicit enough, which allows for a relatively easy use without needing much prior programming knowledge.

PERCEPTORY [BÉD 04] offers a conceptual solution to the issue of multiple representations, with the possibility of describing three-dimensional spatial objects. However, it suffers from a lack of spatial, temporal, and causal notions at an association level. On this point, PERCEPTORY is less powerful than MADS. The pictogram multiplication (along lines, columns, accolades, spaces, etc.) forces spatiotemporal information system designers to a certain learning curve.

In spite of the improvements brought to relational DBMS with conceptual modeling tools, there are still two main disadvantages. First of all, the expressiveness of the pictogram modeling approach reaches its limits very quickly when expressing integrity constraints. MADS is the only tool that allows us to express space and time constraints, but these constraints apply only to association (and not to classes) and there is a very limited choice. Moreover, the easy use aspect of the code generation approach reaches its limits as soon as users wish to apply modifications or changes to their application. They must either regenerate the data diagram (which implies redoing the data acquisition work), or modify the DBMS diagrams by hand, and in that case, we lose the benefits of the conceptual approach. Owing to its expandable AML, AROM can avoid these pitfalls. Therefore, the idea of using it to design and implement applications with spatiotemporal information seemed obvious to us.

4.2.3.2. AROM's *contribution*

Parent *et al.* [PAR 99] list a series of evaluation criteria for spatiotemporal application design tools. Among expected qualities for a good conceptual modeling tool for spatiotemporal information systems (STIS from now on), the following are to be taken into account:

– simplicity;

– conceptualization seen as the direct match between real objects and model entities;

– language's expressive power, in terms of the integration of object types, association types, specialization relations, aggregation, multivalued attributes, integrity constraints, etc.;

– visual convention use for modeling;

– understandability;

– orthogonality between the spatial dimension and the time dimension;

– formal definitions.

By reviewing all the criteria, we realize that for the most part they are all linked to modeling in general, and that the AROM language satisfies them all except for the orthogonality between space and time, since in its basic version, AROM offers neither representations nor processes dedicated to time or space.

Indeed, AROM's language, based on a modeling approach similar to UML's, offers a generic, simple, and accessible representation language. This simplicity is due to an abstraction process of the modeling approach, since the number of typed entities used for the modeling was rather limited. Accessibility is due in part to the similarity between the concepts of the object paradigm and our perception of the concepts and objects in real life; and, on the other hand, to the notoriety of UML language in the digital world.

The expressiveness of the AROM language is due to the fact that it allies object-oriented modeling concepts with the power of an AML. It allows us to create simple models that can be enriched by a wide variety of integrity constraints: typing, field, multiplicity, and cardinality constraints to which integrity constraints described with AML equations can be added. AROM's AML is, in its basic version, similar to OCL [UML 03] and has an equivalent expressiveness [MOI 07].

The AROM language has a visual design language, based on a graphical representation similar to UML's class diagrams. To guarantee the solidity of AROM and the mechanisms it must welcome, its representation language has been provided with denotational semantics based on an abstract syntax of AROM.

These arguments led us to the conclusion that AROM is a favored tool for object and spatiotemporal relation modeling due to classes, associations, and the AML; it could thus be a very interesting solution. However, space and time management were not one of the goals of the system's designers. We have explored the possibility to expand AROM with concepts linked to time and space. There are two possible solutions:

– use the system with its current representation possibility and implant complex abstract data types (ADTs) (polygon, polyline, etc.) such as associations of system base types (for example, points and real number pairs);

– extend the AROM model to introduce the concepts and spatiotemporal types.

AROM has already been used as a design and data management tool for spatiotemporal information systems [BIS 04, MOK 04]. However, previous solutions, based on the use of an ADT, had certain inconveniences linked to the complexity of the time and space data.

ADTs can indeed be created by using the system's modeling possibility. They are then represented as classes covering various attributes of base type or, if necessary, as various classes linked by associations. Using this approach, it is *theoretically* possible to infinitely expand the base types of AROM. This simple and rather efficient solution has however many disadvantages.

From a modeling point of view, ADTs introduce a "pollution" of knowledge bases by introducing "low-level" modeling entities that complicate the model. The need to model and explicitly manipulate ADTs on which the field representation is built, distances the user from its first modeling preoccupations and forces him/her to mix concepts belonging to the modeled field with concepts linked to ADTs. In conclusion, given the advantages of a solution based on the extension of the AROM metamodel and its type system, we have favored it in our approach. In the following section, we will first present the AROM-ST metamodel before detailing the type system extension.

4.3. AROM-ST

4.3.1. *Metamodel*

The AROM-ST metamodel allows us to model independently the dimensions of space and time. The term "independently" here means that the spatial and non-spatial objects must behave in the same way versus time, and, reciprocally, time and non-time objects must behave in the same way versus space. This independence allows, for example, ascendant compatibility of data models: an AROM knowledge base is compatible with the AROM-ST language (and model).

The introduction of temporality and spatiality at a class level is carried out by adding attributes (see Figure 4.1).

Classes situated at the top of the hierarchy represent modeling entities already present in AROM-ST (Object,

`Attribute`, and `Link` classes in Figure 4.1). The spatiotemporal entities extend them by adding space and/or time attributes:

– To change a non-spatial object (`Object` class in Figure 4.1) into a spatial object (`Spatial Object` class), we only need to add a space (type) attribute (`Spatial Attribute` class) that represents its spatial extent.

– To change an atemporal object (sustainable) into a temporal object (`Temporal Object` class), we must add a time (type) attribute (`Temporal Attribute` class) that represents its lifespan.

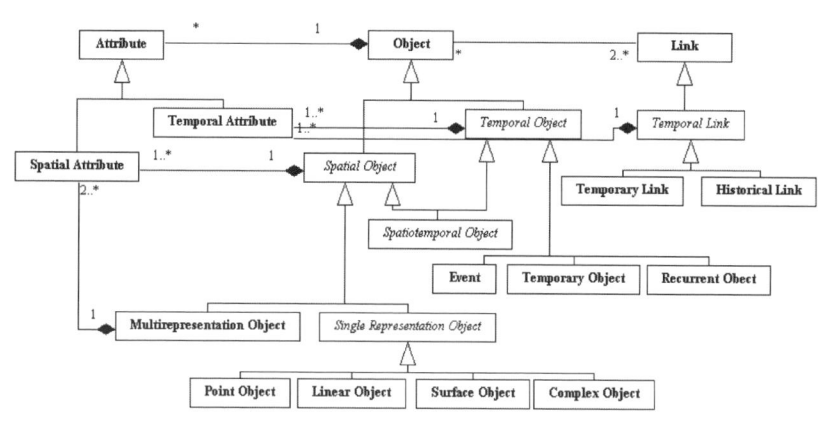

Figure 4.1. *The* AROM-ST *metamodel*

When a spatiotemporal object (`Spatiotemporal Object` class) has both a temporal attribute and a spatial attribute, it inherits the behavior of both the spatial and temporal objects.

Adding spatial attributes creates no conceptual difficulty: the spatial objects can be one-time objects, linear objects, surface objects, or complex objects, depending on the concrete type of the spatial attribute that represents their extent. To take into account spatial objects to which various representations of the spatial extent are associated, we have created, in the AROM-ST model, a dedicated class, the Multirepresentation Object class.

To take into account the display of maps with various scales [PAR 99], we chose to integrate into the AROM-ST model the ability to store various geometries for the same object. Indeed, the multiscale display creates a series of cartographic generalization issues; the same geographical objects must be represented at various scales with different levels of detail:

– For large scales, the geographical objects must be represented with more geometric details.

– For small scales, the fine geometrical details are not visible, but, given that the number of simultaneously visible geographical objects becomes important, it affects application performances.

Thus, it is necessary to simplify even more the geometry of geographical objects as we lower the display scale. Unfortunately, due to its complexity, the cartography generalization process cannot be automatically carried out when the query is launched. The only efficient solution to this issue is to allow for the storage of various geometries for the same geographical object.

4.3.2. *Objects and time relationships*

The temporal dimension can be assessed according to two aspects for real-world objects:

– The existence of the objects itself, which determines their creation/destruction cycle. The temporal aspect is then placed at the level of the object.

– The evolution of the objects during their lifetime, which determines their change of state. In that case, the temporal aspect is placed at the level of the attributes.

AROM-ST's temporal model manages these two aspects.

It is important to note the difference between the existence of the objects in the real world and the existence of the objects in the system. A temporal knowledge base has a historical dimension, meaning that it keeps all the past objects. At the end of the lifecycle of a temporal object, the system records only the fact that the object is not valid anymore, but it is not *physically* deleted from the knowledge base. The temporal objects that are no longer valid can thus be questioned through historical queries.

The introduction of temporality at the level of AROM classes allows us to express the lifecycle of real-world objects. From a conceptual point of view, when it comes to lifecycles, we can consider four kinds of objects: sustainable objects, objects with a limited lifespan, objects with a one-time existence, and recurring objects.

Atemporal objects have no lifespan. These objects can be represented by AROM standard class type. We should note that even if these objects have no temporal dimension at the level of their lifecycle, they can have attributes that may vary in time.

Limited lifespan objects are objects that have their lifecycle (creation/destruction) recorded in the database. These objects are kept at the level of the knowledge base even after their destruction and can thus be questioned due to queries. The classes that represent this type of object have an `Interval Lifespan` attribute that indicates the object's lifespan.

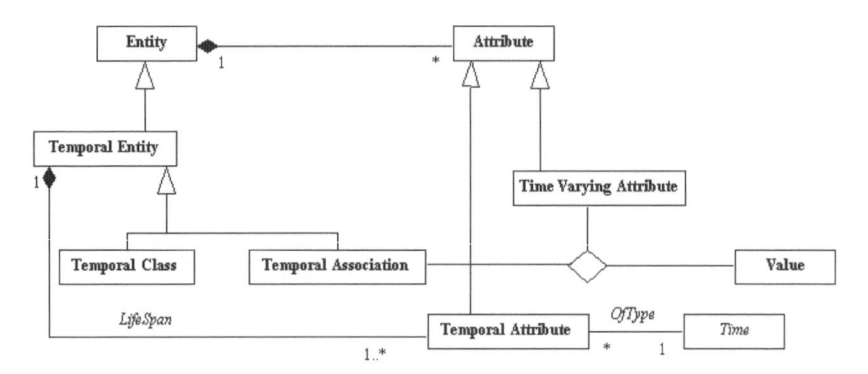

Figure 4.2. *Temporal objects and attributes in* AROM-ST

One-time objects make up events (phenomena that can be natural, social, etc.) that require, due to their importance, to be modeled as such in the knowledge base. Their representation on the time axis is a point. The classes that represent this type of object have an `Instant` type attribute.

Recurring objects are objects whose creation/destruction cycle can be taken up more than once, processes or phenomena that appear and disappear within a certain periodicity. The classes that represent this type of object have a `CompositeTime` type attribute (set of moments or temporal intervals).

Atemporal associations are associations linking atemporal objects. Introducing temporality in the system creates two types of associations: lifespan associations and historical associations.

Lifespan temporal associations describe a relation to the real world that exists as long as the objects involved coexist. For example, it is possible to model a neighborhood situation between two buildings through an association. This association has meaning only during the period in which the two buildings exist; the destruction of a building leads to the destruction of the association. Such an association can link various object classes, among which at least one must be temporal. The lifespan of this type of association cannot go beyond the temporal interval which is a result of the intersection of intervals that represent the lifespans of the objects making up the association. This means that the objects participating in the link must all be valid (existing) for the link to be valid.

The second type of association with a time component described a link that starts when all the associated objects exist, and that persists even after the disappearance of one or more objects creating it. This type of association has a historical dimension. Relations that have a "cause–effect" type of component must be represented by this kind of association.

For example, an association describing the neighborhood relation between two persons is valid only for a limited time, as long as the persons coexist and live in neighboring apartments.

However, an association describing a "cause–effect" type of relation, such as a father–son relation, is a historical type of association. The blood relationship between the two people starts when the son is born and remains true even if one or both persons have ceased to exist in the real world (and are no longer valid within the knowledge base).

In AROM-ST, it is possible to manage the evolution of objects over time, to represent time varying objects (`TimeVaryingAttribute` class in Figure 4.3). To that end, we must draw a line between time varying attributes, which are any type of attribute whose value changes over time, and temporal attributes (`TemporalAttribute` class in Figure 4.3), which have a fixed value of temporal type (`Instant`, `Interval`, `MultiInstant`, or `MultiInterval`).

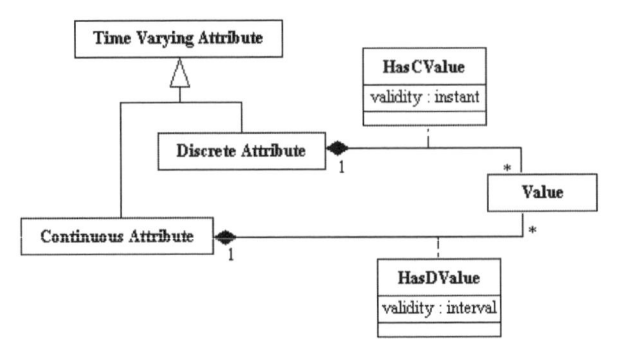

Figure 4.3. *Representation of attributes with discrete and continuous variations*

Time varying attributes can be put into two categories, depending on the type of change they incur:

– attributes whose time variation happens continuously and whose measured value validity is limited to the measuring moment (ContinuousAttribute class in Figure 4.3);

– attributes that incur discrete changes in increments (DiscreteAttribute class in Figure 4.3). They keep the same value during a certain time, but change sharply in value. In that case, the value validity period is described by a temporal interval.

One specific case is that of continuous variation attribute subject to interpolation. From a conceptual point of view, it is unnecessary to overload the model by creating a new type of attribute. However, at the level of implementation, there are more performing storing methods for this type of attribute.

From these measured values, it is possible to extract a function to calculate the value of the attribute versus time, and then to store its parameters. At the time of the query, the value of the attribute is calculated versus time. Storing such attributes is possible in AROM-ST through objects equipped with AML expressions to calculate interpolations. Thus, it is only a more efficient method in case the number of attribute values is too high (real-time systems, for example).

The objects become temporal (or, respectively, spatial) as soon as we add temporal attributes (or, respectively, spatial attributes). A temporal object must have at least one temporal attribute that represents its lifespan. A spatial object must have at least one spatial (geometrical) attribute that gives its geographical representation. Spatial attributes can have geometrical representations for one or more cartographic scales.

4.3.3. *Space and time types*

AROM-ST's spatial types (see Figure 4.4) are in line with the Open Geospatial Consortium (OGC) [OGC 08] specifications. According to this norm, the necessary and sufficient set of types to represent spatial objects is made of the following simple geometrical types: Point, Polyline, Polygon. The Line and LinearRing types are defined by applying constraints to the Polyline type.

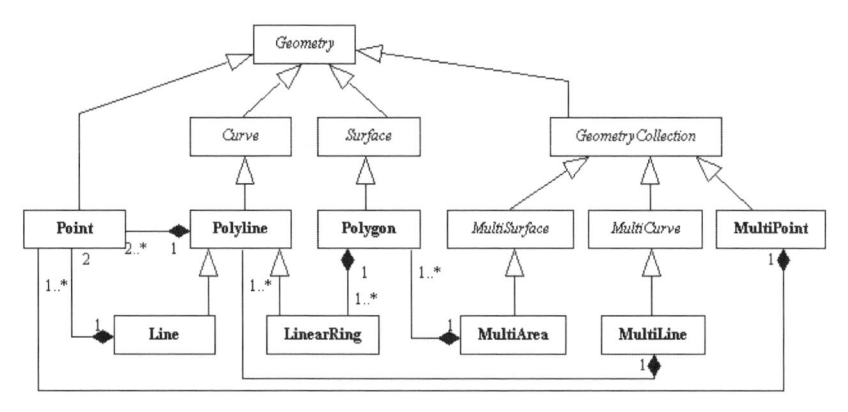

Figure 4.4. AROM-ST's *geometrical model, in line with OGC specifications*

From these simple types, complex (compound) geometrical types can be defined as seen in Figure 4.4: `MultiPoint` (a cloud of points), `MultiLine` (a set of polylines), and `MultiArea` (a set of polygons). The simple temporal types in AROM are the `Instant` type and the `Interval` type. Compound types making a series of instants or intervals are the `MultiInstant` and `MultiInterval` types (see Figure 4.5).

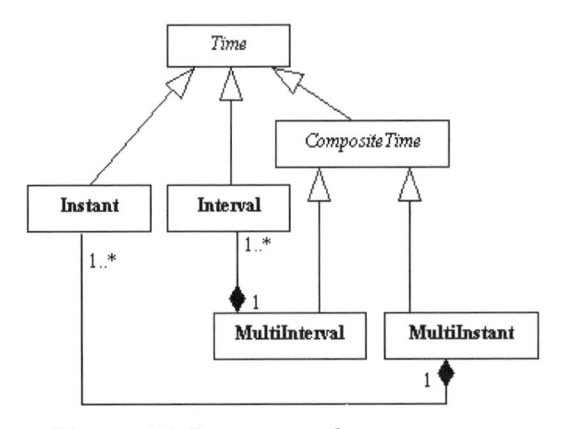

Figure 4.5. *Data temporal type structures*

The available AML operators for spatial and temporal types are divided into three categories: topology operators, set operators, and measure operators. They can be used to carry out quantitative spatiotemporal reasonings (set operators and measure operators) or qualitative reasonings (topology operators). We will detail them in section 4.4 about qualitative and quantitative spatial reasonings.

4.3.4. *Spatial modeling example with* AROM

To illustrate the different aspects of AROM modeling, we are studying a simple example (see Figure 4.6) that allows us to represent the territorial unit hierarchy in a knowledge base dedicated to natural risks.

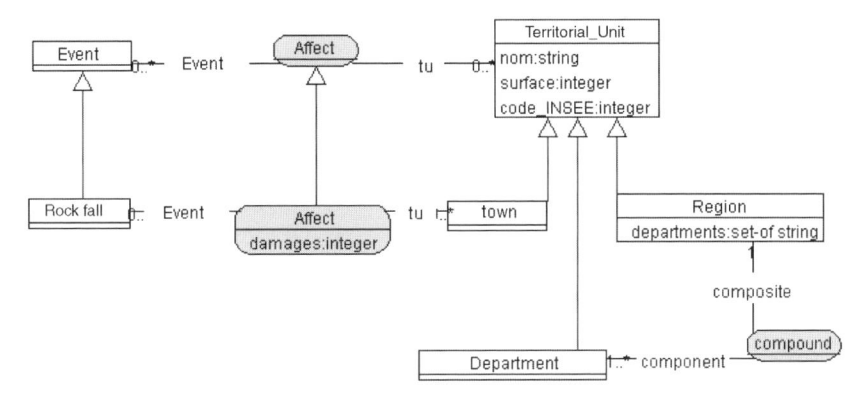

Figure 4.6. *Example of a model designed through the* AROM's *interactive modeling editor (IME)*

A territorial unit (see class Territorial_Unit) can be of two kinds in our example: department or town. The Department class inherits from the Territorial_Unit class but its geometrical attribute contour is redefined by the addition of an AML definition facet that enables us to calculate the geometry of the department by merging the geometry of the outlines of the towns creating it (see the textual format of the base in Table 4.1).

It is interesting to also note that in AROM-ST we can represent n-number of associations between classes (see association result in Figure 4.6).

The territorial units can be affected by different types of natural risk events (class Event in Figure 4.6). The affects association is specialized by the touches association and the addition of a new variable (see the damages variable in Table 4.1). The roles of the association are also specialized, the tu role sees its field restrained to towns, and the ev role sees its field restrained to floods.

```
class: Storm

class: Event

class: Flood
super-class: Event

class: TerritorialUnit
  variables:
    variable: outline
      type: polygon
    variable: area
      documentation:
  << territorial unit
  area >>
      unit: km2

class: Town
super-class:
TerritorialUnit

class: Department
super-class:
TerritorialUnit
  variables:
    variable: outline
      definition: outline =
  merge(this!contains.inf.outline)

association: affects
roles:
  role: ev
    type: Event
    multiplicity:
      min: 0 max: *
  role: tu
    type: TerritorialUnit
    multiplicity:
      min: 0 max: *

association: touches
super-association: affects
```
```
roles:
  role: ev
    type: Crue
  role: tu
    type: Town
variables
  variable: damages
    type: integer
  constraint:
  intersects(this.ev.trajectory,
  this.tu.outline)

association: contains
roles:
role: sup
      type: TerritorialUnit
      multiplicity:
    min: 0 max: *
  role: inf
    type: TerritorialUnit
    multiplicity:
      min: 0 max: 1
  constraint: this.inf.area
< this.sup.area

association: result
  roles:
    role: cause
      type: Storm
      multiplicity:
        min: 0 max: 1
    role: effect
    type: Flood
      multiplicity:
        min: 0 max: *
    role: on
      type: Town
      multiplicity:
        min: 0 max: *
    constraint:
    intersect(this.cause.footprint,
    this.on!a.basin.outline) and
    before (begin(this.cause.period),
    begin(this.effect.period))
```

Table 4.1. *Textual format of the knowledge base presented in Figure 4.6*

The association contains describes the composition relation that exists between a department and various towns. The multiplicity constraints are seen on the roles (sup role with a multiple of 0 or 1, 0 taking into account the case of a higher level territorial unit, which is not the part of other territorial units), and the inf role, which is a multiplicity of 1 to n).

Some modeling details are visible only at the textual format of the knowledge base (see Table 4.1).

In Table 4.1, we can see that the Territorial_Unit class is described by an area variable, which itself is described by a documentation facet and a unit facet representing the measure unit of this area. We can also note the use of integrity constraints on associations to avoid any data input or update error. A constraint on the area value of the territorial units (see the contains association in Table 4.1) allows us to specify that the constituent territorial units cannot have a larger area than the area of their respective constituent units.

We can also note the use of spatiotemporal constraints. A simple spatial constraint enables us to specify the fact that a flood's trajectory that touches a town must intersect the outline of the town. A more complex spatiotemporal constraint allows us to specify that, for a storm to be considered as the cause of a flood, the beginning of the storm must be before the beginning of the flood, and that the storm must have affected the hydrographic basin of the river in which the flood happens. This type of complex spatiotemporal constraint cannot be expressed in "classical" conceptual modeling approaches; we have to express here constraints of mist (spatial and temporal) natures on elements linked by two associations, one ternary and one binary.

4.4. From AROM-OWL to ONTOAST

In more recent work [MIR 07b], we have studied opening AROM to the semantic Web, by trying to bring it closer to OWL-DL [MCG 04], the standard ontological representation language. This comparative study was carried out along three axes: (1) representation, (2) typing, and (3) inferences. After coming to the conclusion that there was an imbalance between AROM-ST and OWL-DL on a description power level, we suggested an extension to its metamodel to close this gap [MIR 07b]. The new metamodel, called AROM-ONTO, integrates new representation structures that bring AROM much closer to OWL-DL. Finally, we positioned AROM in terms of typing and inference versus OWL-DL, by noticing and claiming a certain amount of complementarity.

Thus, with AROM-ONTO, from now on, we can access inferences offered in description logics (which can be carried out by using reasoners such as PELLET[3] and RACERPRO[4]) but it is also possible to enrich the OWL-DL ontologies through AROM-ST reasonings that complement those offers by OWL-compatible reasoners. To benefit inference capacities offered by AROM-ONTO, the ontological knowledge must be translated into AROM's own formalism. To this end, we have built a translator based on XSLT rules that allow us to import an OWL-DL ontology into AROM-ONTO.

AROM-ONTO represents the base from which we have built a spatial and temporal reasoner able to use OWL-DL ontology, called ONTOAST (after ONTOlogy in AROM-ST) and which is a bridge between the GIS field and the semantic Web. ONTOAST is an AROM-ST extension that covers the spatial and temporal operators managed by

---

3 http://clarkparsia.com/pellet/
4 http://www.racer-systems.com/

AROM-ST as well as the core modifications caused by AROM-ONTO to ensure OWL-DL compatibility. While AROM-ONTO guarantees general compatibility between AROM and OWL-DL's ontological languages, ONTOAST offers the expansion of the OWL ↔ AROM-ONTO translators so that they take into account the spatial and temporal descriptions made using GeoRSS-simple and OWL-time ontologies. Thus, the spatial and temporal reasoning offered by ONTOAST can be used to exploit spatial and temporal descriptions defined by OWL-DL ontologies, as long as there is a translation toward and from ONTOAST.

ONTOAST integrates a predefined spatial model that manages a set of qualitative spatial links. These are qualified as topological, distance, or direction links and their respective semantics are defined in [MIR 07a]. By relying on the use of AROM's AML [PAG 01] and on a set of qualitative spatial link deduction rules from other existing qualitative links and digital data [MIR 07a], ONTOAST also manages all Allen's temporal links [ALL 83] (*before*, *after*, *starts/started-by*, *finishes/finished-by*, *during/contains*, *equals*, *meets/met-by*, *overlaps/overlapped-by*). The goal is to answer spatial and temporal queries on "the fly" by combining qualitative reasoning and quantitative reasoning without having precalculated and stored in the ontological base all the possible spatial links between the entities.

4.5. Architecture

Knowledge representation in AROM is implemented through an environment created in the JAVA language. In its current version (AROM 2) [PAG 01], AROM is the implementation of a knowledge representation model: it covers representation in a shape of digital objects of AROM's model entities. AROM 2's API is the JAVA programming interface of this system. Above the core, the AROM 2 platform offers a certain number of libraries for knowledge base use, each of which is built on the AROM 2 API and offers its own API. Among the different modules offers, we can mention AROM CLASSIF, an API for object and propagation tuple classification [CHA 03], AROM QUERY for query definition in an AROM base, and the XAROM API for AROM base interfacing with XML format (Figure 4.7).

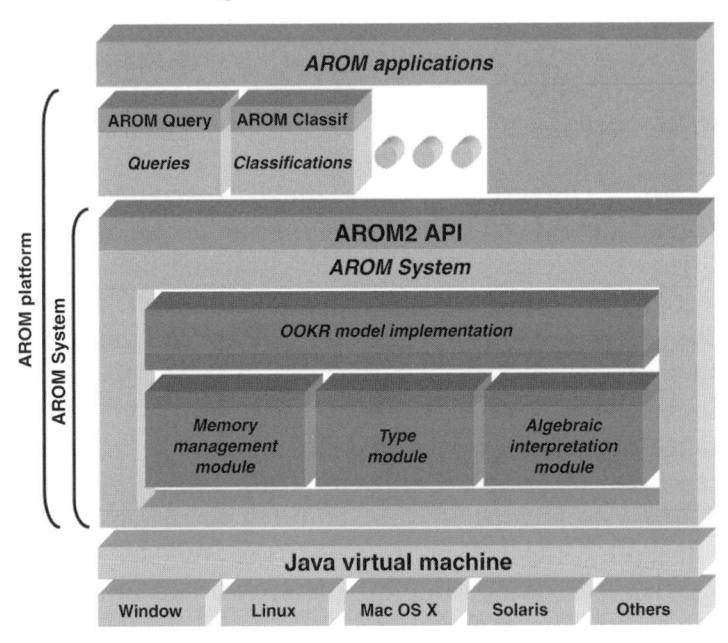

Figure 4.7. AROM *platform architecture*

The AROM system itself is designed in a modular manner, each module being in charge of a specific function in the system. The memory management module ensures memory access to the AROM instances, the type module defines all the recognized types in an AROM base and the possible operations that can be carried out on these types [CAP 98], and the algebraic interpretation module ensures AML equation interpretation in AROM. Each module communicates with the other modules only through a clearly separated internal API. It is thus possible to modify the AROM platform by changing the implementation of a module for another implementation. This possibility has, for example, been used in the GENOSTAR[5] [DUR 03] system that is dedicated to the exploratory analysis of genomic sequences. Since GENOSTAR uses the AROM system for knowledge bases that involve a great number of instances, a specific memory management module allows us to optimize the memory use of the JAVA virtual machine. A persistence mechanism ensures the upload and download of the instances from the disc to the memory. In the same way, a specific type of module integrating genomic specific types is used in this application.

The AROM 2 system configuration happens transparently (through resource files) for the AROM applications. Indeed, AROM 2's API is made of only JAVA interfaces and "fabrication" classes; the AROM's application code never refers to the system's implementation classes. It is thus possible to carry out without modification the same code with different implementations of the AROM system. This new configurable and modular architecture, as well as these new library, is the difference between AROM 2 and the previous version, the semantics themselves have not changed.

4.6. Community

The AROM-ST user community is mostly linked to the need to model data in fields such as bioinformatics and geomatics. AROM uses for bioinformatics mostly take advantage of AROM's classifying abilities. The use of AROM-ST in geomatics, for natural risk modeling or complex system modeling, take advantage of all AROM-ST's innovative characteristics: classification, spatiotemporal types, and constraints expressed in AML.

---

5 http://www.genostar.org/

To illustrate our argument, we would like to take for example the use of AROM-ST for an application called SIHREN dedicated to raising awareness of natural risks and preventing them. The idea underpinning it is to make the modeling steps easier for applications dedicated to natural risks, by providing designers with a generic model devoted to natural risks. This model allows non-specialists (urban area risk planners, hydrologists, etc.) to model their data related to natural risks in a relatively easy manner, by specializing a high-level "natural risk" model (see Figure 4.8).

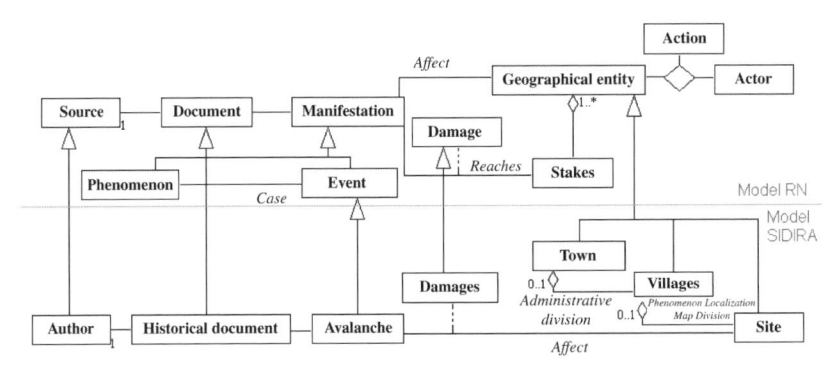

Figure 4.8. *Data model for avalanche risks as specialization of the generic model devoted to natural risks*

This high-level model is based on AROM-ST's spatiotemporal modeling abilities. Once the modeling of the data is finished, the designers have the possibility of visualizing their data in a spatiotemporal interface by importing their data model into the GENGHIS application [MOI 08].

4.7. Conclusions and prospects

In this chapter, we have presented AROM-ST, an OOKRS. AROM combines oriented modeling simplicity with the expressive richness of an AML to offer a modeling and spatiotemporal data management solution that is completely declarative, relatively easy to use, and very expressive.

In fact, AROM-ST has proved itself not only as an independent spatiotemporal data modeling tool, but also as a spatiotemporal data management and modeling tool integrated into the spatiotemporal application generation platform GENGHIS (see Chapter 5).

Future developments for AROM-ST will focus on the problematics linked to the semantic Web. The idea behind this is to bring AROM-ST closer to OWL and to use AROM-ST's spatiotemporal reasoning abilities to fill the gaps left by OWL in the spatiotemporal reasoning field.

4.8. Bibliography

[ALL 83] ALLEN J.F., "Maintaining knowledge about temporal intervals", *Communication of the ACM*, vol. 26, pp. 832–843, 1983.

[BAA 03] BAADER F., CALVANESE D., MCGUINNESS D., NARDI D., PATEL-SCHNEIDER P. (eds), *The Description Logic Handbook: Theory, Implementation and Applications*, Cambridge University Press, Cambridge, United Kingdom, 2003.

[BÉD 04] BÉDARD Y., LARRIVÉE S., PROULX M.-J., NADEAU M., "Modeling geospatial databases with plug-ins for visual languages: a pragmatic approach and the impacts of 16 years of research and experimentations on perceptory", *COMOGIS Workshops ER2004*, LNCS 3289, Shangai, China, pp. 17–30, 2004.

[BIS 04] BISSLER T., Conception et développement d'une plate-forme générique pour les Systèmes d'Information Spatio-Temporelle pour la gestion des Risques Naturels, mémoire d'ingénieur, CNAM Grenoble, 2004.

[CAP 98] CAPPONI C., "Un système de types pour la représentation des connaissances par objets", *Revue d'Intelligence Artificielle*, vol. 12, no. 3, pp. 309–344, 1998.

[CHA 03] CHABALIER J., FICHANT G., CAPPONI C., "La classification récursive en AROM: Application à l'identification de systèmes biologiques", *RSTI, L'Objet*, vol. 9, nos. 1–2, pp. 167–181, 2003.

[DUC 98] DUCOURNAU R., EUZENAT J., MASINI G., NAPOLI A., (eds), *Langages et modèles à objets – Etat des recherches et perspectives*, Collection Didactique D-019, INRIA, Le Chesnay, 1998.

[DUR 03] DURAND P., MÉDIGUE C., MORGAT A., VANDENBROUCK Y., VIARI A., RECHENMANN F., "Integration of data and methods for genome analysis", *Current Opinion in Drug Discovery and Development*, vol. 6, no. 3, pp. 346–352, 2003.

[MCG 04] MCGUINNESS D.L., VAN HARMELEN F., OWL Web Ontology Language – Overview, W3C Recommendation, World Wide Web Consortium, 2004, http://www.w3.org/TR/owl-features/.

[MIN 75] MINSKY M., "A framework for representing knowledge", *Reading in Knowledge Representation*, McGraw-Hill, New York, pp. 245–262, 1975.

[MIR 07a] MIRON A., CAPPONI C., GENSEL J., VILLANOVA-OLIVER M., ZIÉBELIN D., GENOUD P., "Rapprocher AROM de OWL", *Langages et Modèles à Objets*, Toulouse, France, 2007.

[MIR 07b] MIRON A., GENSEL J., VILLANOVA-OLIVER M., MARTIN H., "Towards the geo-spatial querying of the semantic web with ONTOAST", *7th International Symposium on Web and Wireless GIS (W2GIS 2007)*, Cardiff, UK, 2007.

[MOI 07] MOISUC B., CAPPONI C., GENOUD P., GENSEL J., ZIÉBELIN D., "Modélisation algébrique et représentation de connaissances par objets en AROM", *Langages et Modèles à Objets*, Toulouse, France, 2007.

[MOI 08] MOISUC B., GENSEL J., MARTIN H., "Conception de systèmes d'information spatio-temporelle adaptatifs avec ASTIS", *Numéro spécial de la Revue Nouvelle des Technologies de l'Information*, Cépaduès, RNTI E-13, pp. 63–78, 2008.

[MOK 04] MOKRANE A., LAOUAMRI O., DRAY G., PONCELET P., "Modélisation spatio-temporelle des connaissances d'un système d'information géographique", *Proceedings of SETIT'04 International Conference*, Sousse, Tunisia, 2004.

[NAP 04] NAPOLI A., CARRÉ B., DUCOURNAU R., EUZENAT J., RECHENMANN F., "Objets et représentation, un couple en devenir", *RSTI L'objet*, vol. 10, no. 4, pp. 61–81, 2004.

[OGC 08] Open Geospatial Consortium, OGC Reference Model, 2008, http://www.opengeospatial.org/standards/.

[PAG 00] PAGE M., GENSEL J., CAPPONI C., BRULEY C., GENOUD P., ZIÉBELIN D., "Représentation de connaissances au moyen de classes et d'associations: le système AROM", *Actes de la conférence LMO 2000*, Montréal, Canada, Hermès-Lavoisier, pp. 91–106, January 2000.

[PAG 01] PAGE M., GENSEL J., CAPPONI C., BRULEY C., GENOUD P., ZIÉBELIN D., BARDOU D., DUPIERRIS V., "A new approach in object-based knowledge representation: the AROM system", *Engineering of Intelligent Systems*, LNCS 2070, Springer, Berlin, pp. 113–118, 2001.

[PAR 99] PARENT C., SPACCAPIETRA S., ZIMÁNYI E., "Spatio-temporal conceptual models: data structures + space + time", *Proceedings of the 7th ACM Symposium on Advances in GIS*, Kansas City, Kansas, USA, 1999.

[SPA 07] SPACCAPIETRA S., PARENT C., ZIMÁNYI E., "Spatio-temporal and multi-representation modeling: a contribution to active conceptual modeling", *Active Conceptual Modeling for Learning*, LNCS 4512, Springer, pp. 194–205, 2007.

[UML 03] Object Management Group, UML 2.0 OCL Specification, Object Constraint Language, 2003, http://www.omg.org/.

Chapter 5

GENGHIS: an Environment for the Generation of Spatiotemporal Visualization Interfaces

5.1. Introduction

The integration of historical and temporal data in geographical information systems creates issues of methodological and technological natures. Creating spatiotemporal information systems (STIS) requires us to go through design, modeling, and implementation stages that require specific developments. These can be major and time-consuming developments, and even require the intervention of computer specialists. To this day, there are few IT platforms able to help a designer through the different stages of STIS development, from the modeling to the generation of the STIS, while taking into account the geovisualization interface specifications to be implemented.

Chapter written by Paule-Annick DAVOINE, Bogdan MOISUC and Jérôme GENSEL.

The GENGHIS platform (Generator of Geographical and Historical Information Systems), developed by the STEAMER team of the Grenoble Computer Science Laboratory, aims to offer a solution to this issue. The goal is to offer a computer environment able to help the designer during the modeling and creation of an STIS which is adapted to his/her needs. GENGHIS is in line with an approach capitalizing on various research works that have been previously carried out by the STEAMER team on the design and creation of STIS devoted to the representation of spatiotemporal phenomena.

After describing the context in which GENGHIS has been developed (section 5.2), we will present its main functionalities (sections 5.3 and 5.4) as well as its architectural aspects (section 5.5) and then present its scope and specify the targeted user community (section 5.6).

5.2. Context

With the accumulation of data over time, the evolution of information collection processes, or the complexity of issues of social or environmental phenomena, spatial analysis increasingly requires the integration of the temporal or historical dimension in geographical information systems. GIS also cause us to wonder how to visually present and view the multidimensional data. This involves the development of information system applications devoted to spatiotemporal and historical information.

SPHERE and SIDIRA are projects supported by the European Union, the Rhône-Alpes region, the Grenoble center for natural risks (SIHREN) [DAV 06], and the Department for Environment, Energy, Sustainable Development, and the Sea (SIDIRA) and are in line with this line of reasoning. Both aim to design and develop an information system devoted to capitalizing and building on past catastrophic events as well as transmit them to different actors. These events, be they avalanches, landslides, floods, or earthquakes, can be broken down into a phenomenological description, a spatial description or location, a date, and a period or return cycle. Thus, the characteristics of the information manipulated in our projects linked to natural risk management led us to suggest tools able to take into account the different dimensions into which this information could be broken down. To be precise, two specialized applications were originally developed:

– SPHERE: an information system devoted to using historical information on floods [DAV 04];

– SIDIRA: an information system to disseminate avalanche risk information [DAV 03].

We will now present the specificities and appeal of the SPHERE and SIDIRA applications before showing the principles on which the GENGHIS tool is based.

5.2.1. *The* SPHERE *and* SIDIRA *applications: two applications devoted to visualizing data linked to natural risks*

The SPHERE [DAV 04] and SIDIRA [DAV 03] projects showed the necessity of having a tool for information system allowing us to easily cover all the data linked to past events through queries of a temporal (spontaneous data or data related to a specific period of time), spatial (data linked to a geographic entity), thematic (thematic attribute), or documentary (bibliographical, documents, photographies, etc.) nature.

The approach taken to design these tools relies on the principle of visualizing information through a multidimensional interface made of three interconnected frames. Figure 5.1 shows the interfaces developed for the SPHERE and SIDIRA projects.

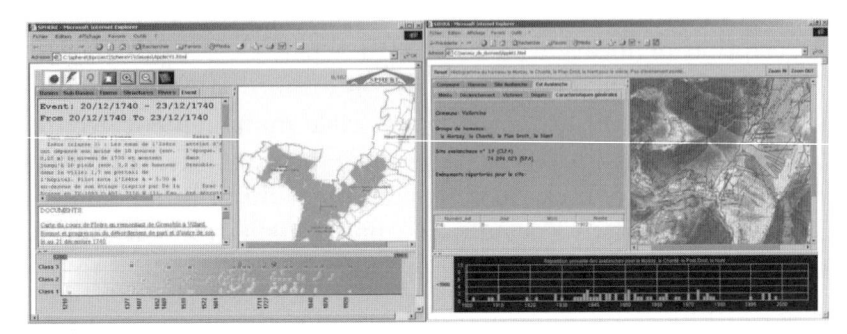

Figure 5.1. SPHERE *and* SIDIRA *application interfaces*

The interface offered is interactive, and allows the user to formulate visual queries with a mouse, for example, and see the result. Each action carried out by the user on one of the application frames then provokes two successive reactions from the system: (i) the query is interpreted and executed and (ii) the results are reflected on all the frames in a synchronized manner:

– any spatial frame allowing us to visualize object spatial attributes (spatial extent, location, trajectory, etc.) through vectorial or *raster* maps and to carry out spatial queries;

– any temporal frame allowing us to visualize object temporal attributes (dating, lifespan, etc.) through temporal diagrams and carry out temporal queries;

– any attributary – or informational – frame allowing us to visualize object attributes through tables and charts, and carry out requests focusing on these elements. It also allows us to display available multimedia documents linked to the selected information as hyperlinks.

The multidimensional visualization concept offered through these experiments appeared as one of the strong characteristics of information systems since they must integrate data with spatial and temporal components. Indeed, this type of visualization to contextualize selected information by replacing it in spatial, temporal, or informational context, has to take its complexity into account. It thus appeared useful to generalize the principles on which the SPHERE and SIDIRA tools were built to other STIS-type applications. This led us to design a generic application allowing the user to design and develop his/her own geovisualization interface by means of models and specific tools.

5.2.2. GENGHIS: *a generator of geovisualization applications devoted to multi-dimensional environmental data*

We can define an STIS along two major characteristics:

– the first linked to content management: the STIS design tools must offer support for spatiotemporal and attributary data modeling, management, and manipulation, without requiring programming efforts on the designers' part;

– the second linked to presentation management: the STIS must allow for the display of spatial, temporal, or attributary data in a dynamic and interactive way, with multiscale and multidimensional navigation possibilities within the informational space of the STIS.

GENGHIS covers all the stages related to STIS development, i.e. modeling and integrating data into the STIS, as well as geovisualization interface modeling and generating. GENGHIS thus relies on two core principles:

– The first is linked to visualizing and viewing the data integrated into the application. GENGHIS has an interface generator, designed to automatically process, from a minimum configuration, the data integrated in the data model. The generated interfaces integrate the visualization and viewing principles developed in the specialized applications SPHERE and SIDIRA and mentioned again in Figure 5.2.

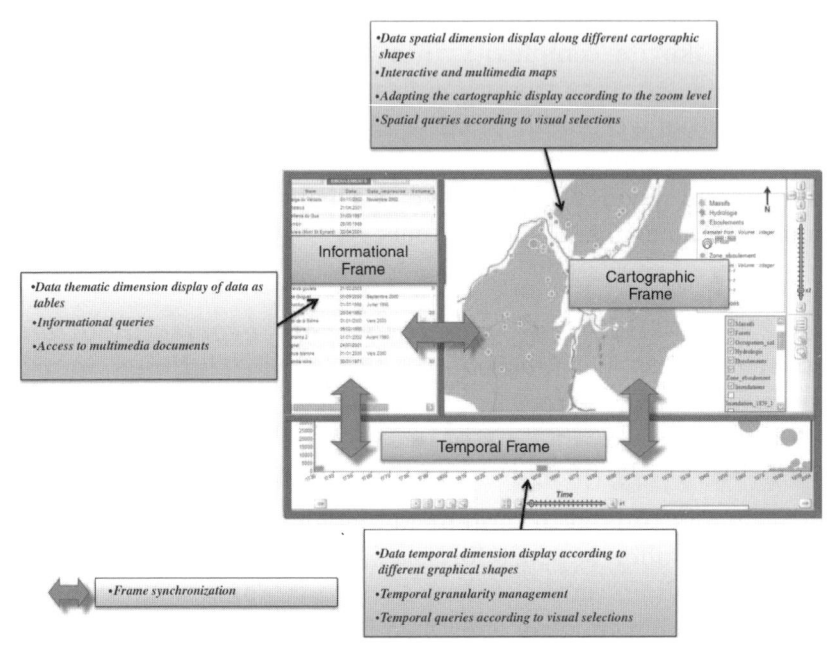

Figure 5.2. *Visualization interface principles in* GENGHIS

– The second is linked to the selected modeling approach. It relies on the distinction between "data model" and "presentation model", and on the use of AROM-ST, an object-oriented knowledge representation system, integrating spatiotemporal representation possibilities [MOI 07]. The presentation model and the data model are closely linked since each data model class is associated with a presentation model class (see section 5.3.3):

- the data model describes, as classes and associations, the real-world objects belonging to the field of studies of the application to be generated, as well as the links between them. Each generated application relies on the definition of a model built from AROM-ST's spatiotemporal metamodel, described in Chapter 4.

- we call "presentation" the way in which the spatiotemporal data are visually represented (graphically and cartographically). The presentation model describes, by means of classes and associations, the way spatiotemporal data of the application domain are to be displayed.

Thus GENGHIS allows us to generate a software application, enabling the user to, on the one hand, visualize multidimensional data contained in a knowledge base, and, on the other hand, query these data according to spatial, temporal, and thematic queries. To this end, two major functionalities are offered [GAY 09]:

– those linked to the creation and generation of the specialized visualization application (section 5.3);

– those linked to the specialized application use, that is the visualization and querying of spatiotemporal data (see section 5.4).

5.3. Functionalities linked to the generation of geovisualization applications

GENGHIS has an interface generator, designed to automatically process, from a minimum configuration, the data integrated in the data model.

5.3.1. *Use cases for* GENGHIS

Figure 5.3 describes use cases offered by GENGHIS when it comes to the generation process of a specialized visualization application. Three main stages make up the creation of a specialized application.

– The first is instancing the data model on which the specialized application is based.

– The second is instancing the presentation model by specifying the different elements that will structure the interface: defining the frames structuring the interface, the geographical and temporal layers matching the modeled entities, and the graphical styles linked to each layer.

– The third stage is launching the application generation process.

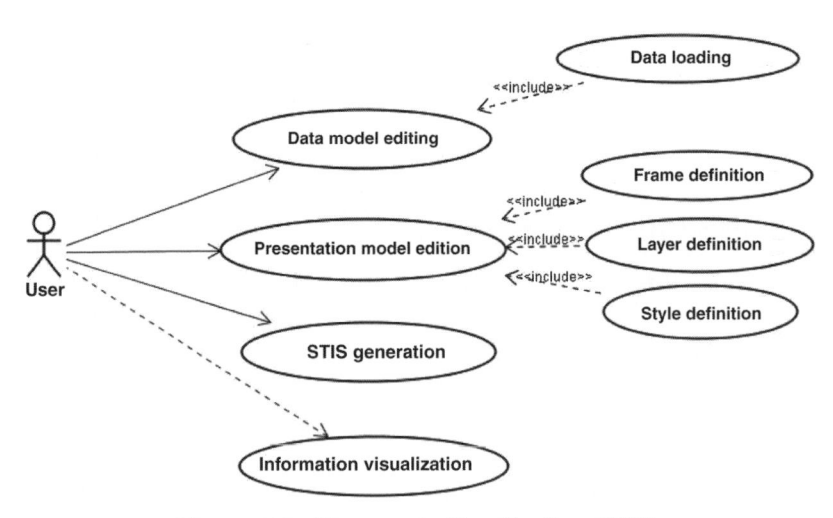

Figure 5.3. *(Use cases) offered by* GENGHIS

A *wizard* helps guide the user during the different stages of the application generation.

5.3.2. *Instancing the data model and the knowledge base*

This functionality allows us to instance the data model which must first be defined in AROM-ST, as well as to populate the knowledge base with objects (class instances) and tuples (association instances) describing the application field. These objects and tuples are created from external data files in XLS, MIF/MID, or *shapefile* format.

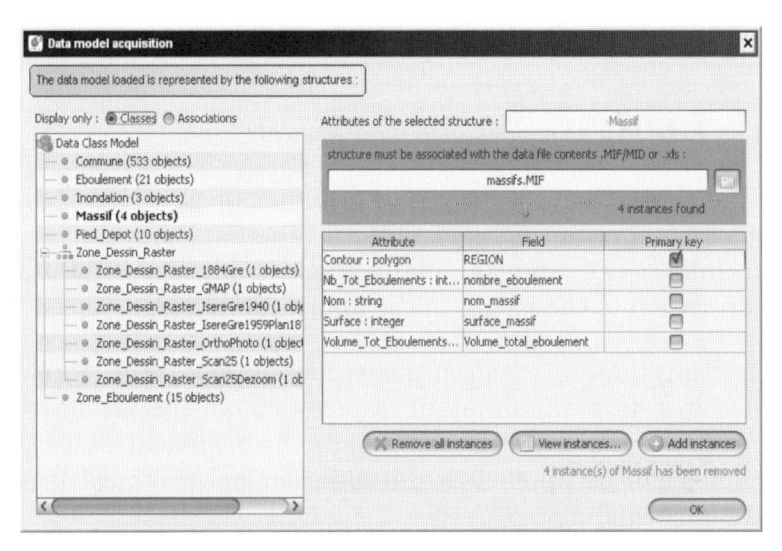

Figure 5.4. *Graphical interface for the instantiation of the data model*

The instantiation of the data model relies on a JAVA interface that works as a controller for an object-relational mapping (ORM). This ORM enables the conversion of stored data according to a relation paradigm – in which the identity of the entity is placed at the level of the attribute value (notion of "key") – and an object paradigm – in which the identity of an entity is generated by the system (notion of object identifier). This prevents there being various entities with identical attribute values for the attributes making up the keys, for example. Figure 5.4 shows an example of mapping creation between the Massif class attributes and the field values placed in an MIF/MID file. We must specify the key for the MIF/MID files to avoid adding double entries in the knowledge base when going through multiple updates.

We should also take into account the differences between the different STIS data models.

To update the knowledge base, we can add, delete, or modify the instances interactively.

5.3.3. *Editing the presentation model*

The goal is to allow the STIS designer to specify graphical elements representing each object of the data model (classes and instances), and to adapt the visualization interface organization to the issues in the application field.

Editing the presentation model means specifying, through a user interface, the different frames to be displayed (spatial, temporal, and attributary) as well as the graphical styles and semiology to be applied to the different geographical layers and temporal diagrams.

From a conceptual point of view, the presentation model is classified according to three types of elements organized in a relative hierarchy: styles, layers, and frames.

– Styles are elementary graphical elements of final presentation that cannot be broken down into simpler elements. A style allows us to specify visual characteristics (color and line thickness, fill color, graphical stakeout, etc.) that the objects to be drawn must inherit. A style is tied to a layer.

– A layer is tied to a data model class. A layer can be geographical or temporal. A layer is defined by the graphical elements and styles that match it.

– Frames are higher level elements. A frame has one or more layers. It can be temporal, spatial, or attributary and has a set of specific functionalities related to its use (legend, layer management, zoom, etc.).

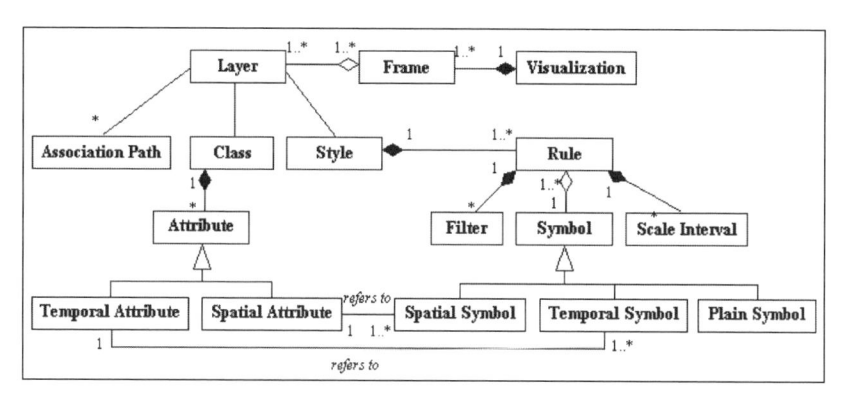

Figure 5.5. *Simplified presentation model of* GENGHIS *[MOI 07]*

The presentation model of GENGHIS takes up the styled layer descriptor (SLD; [OGC 02]) of the Open Geospatial Consortium (OGC) and expands it on various aspects. SLD is a presentation standard that allows us to describe style transformations (Style class in Figure 5.5) to be applied on data layers (Layer class). A data layer refers to a spatial object class (Class class in Figure 5.5, also called FeatureType in OGC terminology). A style is made up of a set of rules that allow us to describe the exact type of graphical symbols corresponding to each spatial object according to its attribute values (Filter class). For example, for a layer representing rivers whose graphical symbol is a blue line, a set of rules allows us to specify a different line thickness according to each river's flow. A rule can also specify the scale interval for which a set of objects contained on the map is visible.

The presentation model for GENGHIS expands the SLD standard on two aspects: (i) taking into account temporal and attributary presentation and not just spatial presentations; and (ii) taking into account dynamic elements of presentation to allow interactive presentations. GENGHIS allows us to define complex and multidimensional visualizations (Visualization class) containing various frames (Frame class) of a spatial, temporal, or attributary type, each frame displaying a certain number of layers (spatial, temporal, or attributary). New temporal (1D, `Temporal Symbol` class) and attributary (0D, `Plain Symbol` class) graphical symbol types were added to existing SLD types (2D, `Spatial Symbol` class).

To enable the dynamic and interactive presentation generation, GENGHIS's presentation model expands the notion of layer described in SLD by adding information concerning existing links between different object types (`Association Path`).

5.3.4. *Generating the geovisualization interface*

Generating the geovisualization interface does not require interaction and information exchange with the user. This stage happens automatically once the data and presentation models are instanced. This functionality operates as the following:

– Generating `SVG` format files corresponding to the spatial and temporal frames specified during the presentation model edition; each file has a set of *vector* and/or *raster* layers and a legend describing the set of layers.

– Generating `XHTML` format files corresponding to the informational frame and containing thematic data.

– Generating specified layers; this generation happens by embedding the graphical elements, at the level of layers it creates the link between graphical elements and styles that they match.

– Generating a *generate* folder containing all the files that make the application (see Figure 5.6).

Nom	Date de modification	Taille	Type
Carte.svg	15 octobre 2009, 15:55	60 Ko	Document SVG
CarteControl.svg	15 octobre 2009, 15:55	12 Ko	Document SVG
CarteLayer.html	15 octobre 2009, 15:55	4 Ko	HTML Document
Eboulement.html	15 octobre 2009, 15:55	16 Ko	HTML Document
GarphTemp_Inondation.svg	15 octobre 2009, 15:55	24 Ko	Document SVG
GarphTemp_InondationGranularity.html	15 octobre 2009, 15:55	4 Ko	HTML Document
GarphTemp_InondationLayer.html	15 octobre 2009, 15:55	4 Ko	HTML Document
GraphTemp_Eboulement.svg	15 octobre 2009, 15:55	32 Ko	Document SVG
GraphTemp_EboulementGranularity.html	15 octobre 2009, 15:55	4 Ko	HTML Document
GraphTemp_EboulementLayer.html	15 octobre 2009, 15:55	4 Ko	HTML Document
images	15 octobre 2009, 15:55	--	Dossier
index.html	15 octobre 2009, 15:55	4 Ko	HTML Document
indexCarte.html	15 octobre 2009, 15:55	4 Ko	HTML Document
indexGarphTemp_Inondation.html	15 octobre 2009, 15:55	4 Ko	HTML Document
indexGraphTemp_Eboulement.html	15 octobre 2009, 15:55	4 Ko	HTML Document
indexTable.html	15 octobre 2009, 15:55	4 Ko	HTML Document
Inondation.html	15 octobre 2009, 15:55	8 Ko	HTML Document
script	15 octobre 2009, 15:55	--	Dossier
Table.html	15 octobre 2009, 15:55	4 Ko	HTML Document
Zone Eboulement.html	15 octobre 2009, 15:55	12 Ko	HTML Document

Figure 5.6. *Set of* HTML *and* SVG *files generated and contained in the generate folder*

The geovisualization interface is made operational by activating, due to an Internet navigator (Firefox, Internet Explorer, Opera, etc.), the index.html file contained in the *generate* folder of the application (see Figure 5.6).

5.4. Functionalities of the geovisualization application generated by GENGHIS

The application generated by GENGHIS is the software component which should allow us:

– to display presentation files (SVG and XHTML files), and thus the visualization interface made of spatial, temporal, and informational frames, as well as of their content;

– to carry out data querying and viewing functionalities;

– to carry out navigation functionalities (zoom, *pan*, layer management);

134   Innovative Software Development in GIS

– to carry out synchronization functionalities between different frames.

Figure 5.7 shows an example of a generated geovisualization interface.

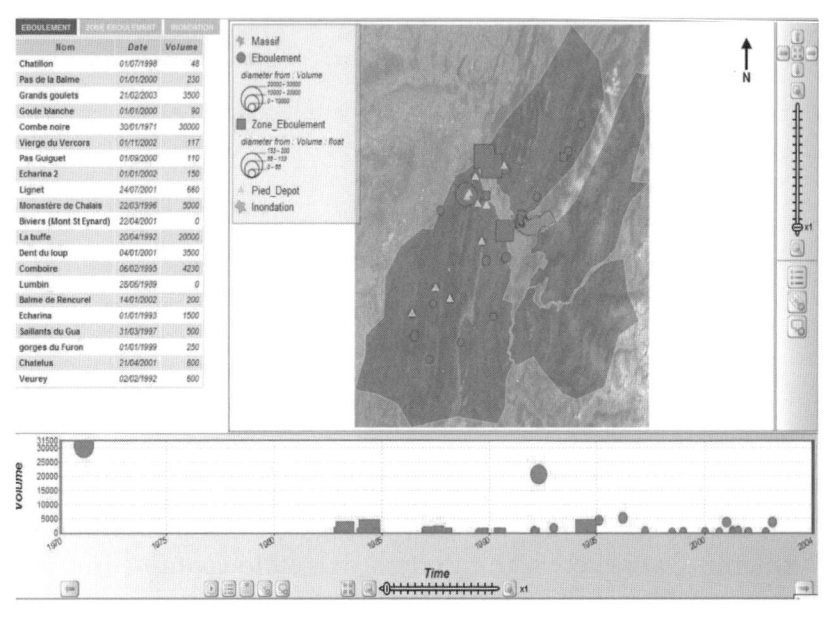

Figure 5.7. *Example of a geovisualization interface generated in* GENGHIS

The interfaces generated in GENGHIS integrate the visualization and viewing principles developed in the specialized applications SPHERE and SIDIRA. The data can be visualized through various synchronized frames, each representing a dimension of the manipulated information: a cartographic frame is devoted to the spatial dimension; a frame integrating the graphics represents the temporal dimension; and an informational frame visualizes the details of each entity contained in the information system as tables. These frames can also integrate a documentary module that allows us to access multimedia content about these entities (texts, images, and maps). Viewing is carried out as a visual request by clicking on the mouse. We will detail now the functionalities of the generated interface.

5.4.1. *Spatial frame functionalities*

The spatial frame (Figure 5.7) displays the interactive maps in an SVG format and offers different control components to ensure the following functionalities:

– management and interactive and dynamic display of the different spatial layers;

– interactive and dynamic management of the legend and tooltips;

– zoom level and layer display management according to a zoom factor;

– *pan* and relocation.

A visual selection on the map, carried out due to a click on the mouse, corresponds to queries of the following type:

– What is the thematic information or data matching the selected spatial object?

– In which period or at which date did the selected object happen?

The result of these queries is then simultaneously highlighted in the thematic and temporal frames.

5.4.2. *Temporal frame functionalities*

The temporal frame (Figure 5.8) is presented as a graph with a right-handed Cartesian coordinate system whose

x-axis is a scale of time and y-axis is a qualitative or quantitative attribute defined by the designer during the data and presentation model creation. This frame offers different control components allowing us:

– to display data on different levels of temporal granularity;

– to select time intervals, thanks to sliders;

– to dynamically and interactively manage the legend, time zoom, and tooltips;

– to manage the temporal animations.

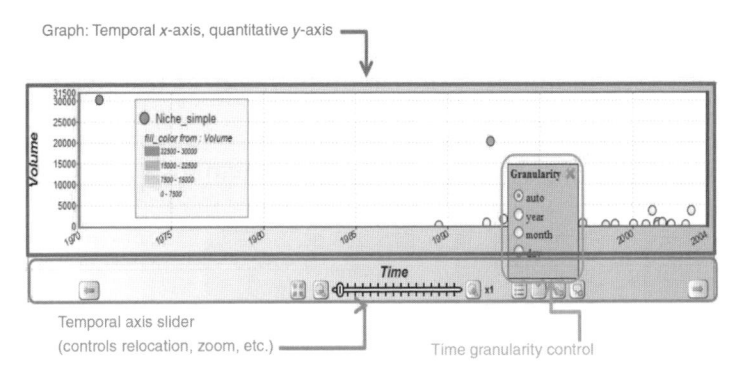

Figure 5.8. *Temporal frame functionalities*

A visual selection on the temporal diagram, carried out due to a click on the mouse, corresponds to queries of the following type:

– What is the thematic information or data matching the objects (or phenomena), which happened at the selected date or during the selected period of time?

– Where are the objects (or phenomena), which happened at the selected date or during the selected time lapse, situated?

The result of these queries is then simultaneously highlighted in the spatial and thematic frames.

5.4.3. *Informational frame functionalities*

The goal of the informational frame (Figure 5.9) is to display the thematic component of the manipulated information. It is displayed as a tab structure called *tabs*. Each tab matches a data model class and is presented as a data table representing class instances. The table columns correspond to the class attributes and instance lines themselves.

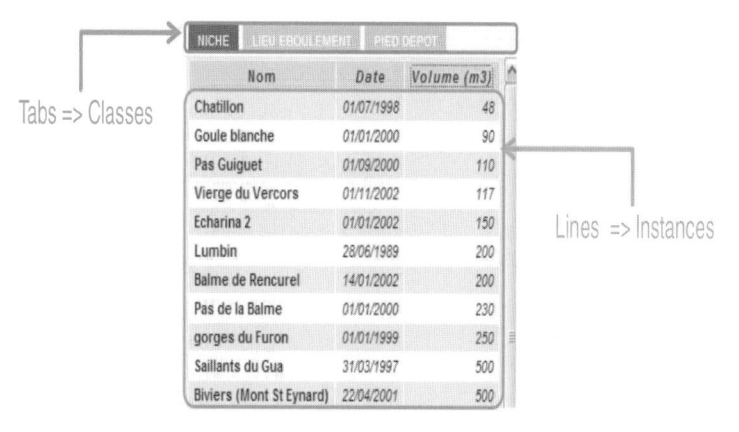

Figure 5.9. *Informational frame*

The attributary part is made of a master HTML document that realizes the tab structure. There is no HTML-specific tag to represent the *tab* component. An in-depth use of tags such as <*li*> (chip management), <*div*> (container management), <*iFrame*> (external file management), and CSS styles allows us to simulate tab behavior. Selecting a tab leads to the display of a different HTML document that describes the data table.

The visual selection, due to a click on the mouse on one of the table's lines, matches the following queries:

– Where is the selected object?

– When did the selected object happen?

138   Innovative Software Development in GIS

The result of these queries then simultaneously highlights the matching objects in the spatial and temporal frames. We can display multimedia documents attached to the objects due to this frame. Figure 5.10 shows pictures characterizing a selected flood.

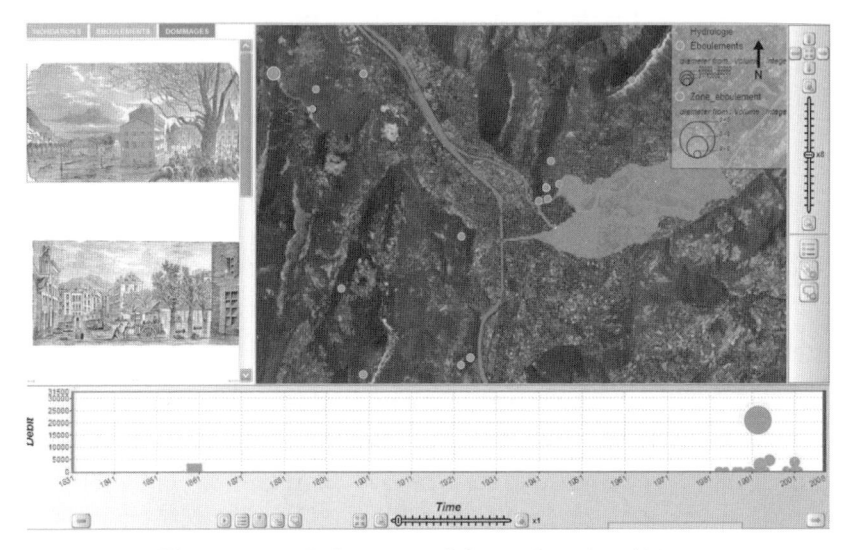

Figure 5.10. *Informational frame functionalities*

5.4.4. *Interactivity and synchronization principles*

The interaction manager is the component, on the one hand, managing the interaction with the user and, on the other hand, synchronizing the visual components. It interprets the user's actions and enables the navigation in the knowledge base informational space. Each user action is interpreted as a query on the knowledge base. The generic query is, in OQL [CLU 98], as follows:

Select O from C where SR and TR

in which:

– *O* represents objects to be displayed in one of the frames of the application;

– *C* represents the object class to be displayed;

– *SR* and *TR*, respectively, represent spatial restrictions and temporal restrictions to be applied to the query.

Each action on the spatial frame modifies the spatial restriction applied to the query. The spatial restriction is defined by the spatial frame coordinates when nothing is selected or by the object's spatial extent when a spatial object has been selected.

Each action on the temporal frame modifies the temporal restriction applied to the query. The temporal restriction is defined by the temporal interval in the temporal frame when nothing is selected, or by the selected moment.

An action on one component determines an update for other components, as shown in Figure 5.11. For example, repositioning the spatial frame (*panning*) causes an update of the temporal frame, now displaying the events affecting the visible spatial entities. A zoom in on the time axis diminishes the number of events displayed in the informational and spatial frames.

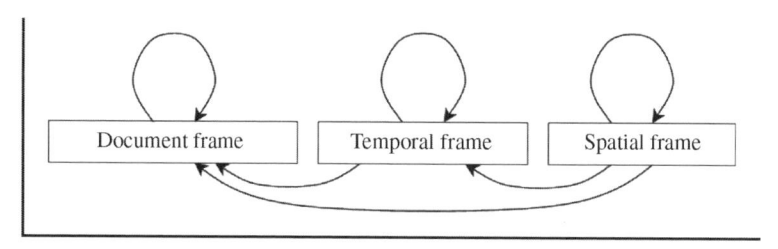

Figure 5.11. *Component synchronization in* GENGHIS

5.5. Architecture

On an architectural level, GENGHIS is made up of two distinct applications:

– A "main" application, made of three modules, is in charge of the required different stages needed for the specialized STIS creation;

– A light-client application, embedded in the STIS, to use the system created with an Internet navigator, and to view and visualize the data.

Figure 5.12 represents GENGHIS's architecture and the software building blocks used.

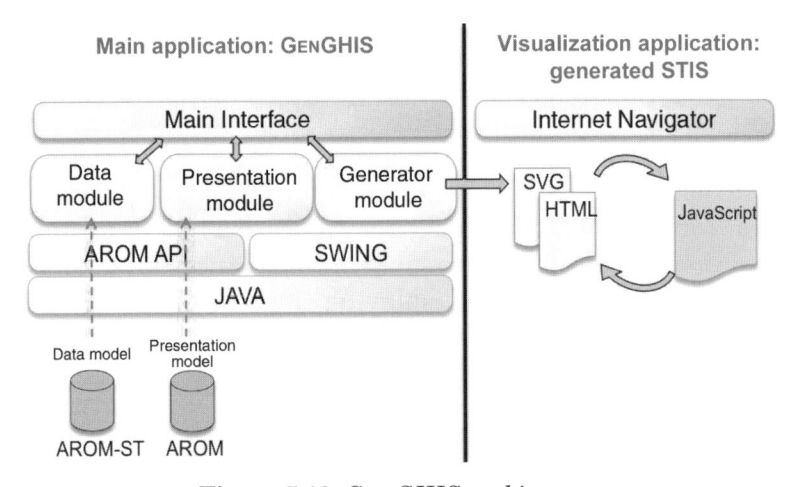

Figure 5.12. GENGHIS *architecture*

The main application is a thick-client application developed in JAVA with SWING-type graphical components. It is made of a main interface meant to guide the user through a succession of different stages, from the design to the generation of an STIS. To that end, three specific modules are offered, each covering the different stages of design:

– one module is in charge of enriching the information in the data model;

– one module enables the user to instance the presentation model consistently with the data model objects;

– one module generates the STIS from the two previously created models. The generator module then builds a structure and classified set of files in HTML and SVG formats. The file data respect a precise tree structure so as to be used and modified by the second application.

The data are stored in the AROM-ST knowledge base. The JAVA interface ensures the editing and instantiation of the data and presentation models, and it is made of a JTREE graphical component allowing us to display as a tree structure, the different model elements (Figure 5.4).

The second application is made of a set of HTML/SVG files created by the first application, but also by JAVASCRIPT files. These script files are the applicative and dynamic part of the second application. We remind the reader here that SVG language was used to represent *vector* forms of graphical and cartographic components, while the HTML language was created for the informational frame display.

5.6. Scope and user communities

5.6.1. *Natural risks: a privileged scope*

GENGHIS is an experimental platform that stems from research and is devoted to the management and viewing of spatially and temporally referenced data. Although it has not yet been released, it has however been involved in different projects focusing on natural risk assessment and management in Grenoble and its suburbs. These projects helped contribute to its development and improvement: SIHREN for the knowledge of past landslides and floods [ARN 09], SIRSEG for building on data linked to the assessment of earthquake risks [CAR 09], and MOVISS [BEC 09] for the spatiotemporal assessment of the social vulnerability versus earthquake risks. These projects are in line with a multidisciplinary approach and are based on the construction of a tool allowing us to gather, structure, organize, and visualize data linked to hazard and/or vulnerability characterization.

Each project required the conception of a data model based on AROM-ST's spatiotemporal meta-model, from which the designer could create his/her own knowledge base and geovisualization interface according to the data characteristics and goals that the final user had. We will now present two application examples.

5.6.1.1. *The* SIHREN *application*

The SIHREN project [DAV 06, ARN 09] is funded by the Rhône-Alpes region and the Grenoble center for natural risks. Its goal is to design and create an information system devoted to building on the characterization of data linked to flood and landslide phenomena that took place in the Grenoble region in the past. This project is carried out in partnership with the Lyon CEMAGREF, the LIRIGM (Grenoble), and the Acthys company. It is in line with the work started in SPHERE and SIDIRA and concerns similar data, which means it integrates a spatial, temporal, attributary, and documentary dimension.

The SIHREN application relies on the design of a "Natural Risk" data model (Figure 5.13) structured around the notion of an event (Event class). These events can be floods or landslides caused by the phenomena described in the Phenomenon class. These phenomena are considered to be natural manifestations (Manifestation class) and affect geographic entities (Geographic Entity). An event can damage different stakes (Stake class). Natural manifestation can be described through documentary content (Document class) for which the source has been identified (Source class).

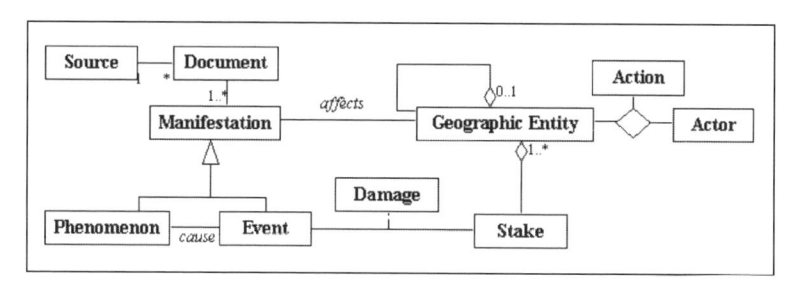

Figure 5.13. *The "natural risk" model underpinning the* SIHREN *application*

Once the data and presentation models are instanced and the knowledge bases specific to landslides and floods populated, the SIHREN application has been generated (see Figure 5.14).

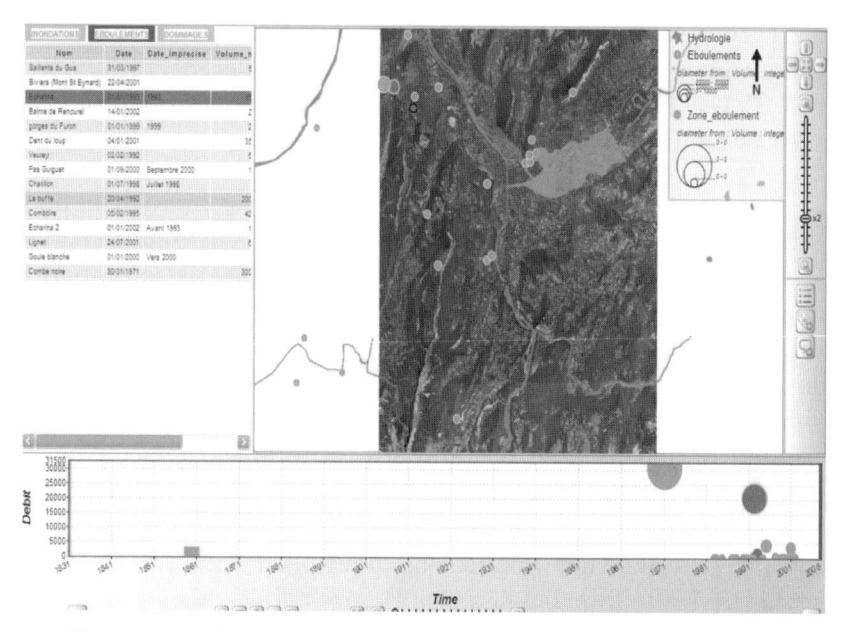

Figure 5.14. *Geovisualization interface of the* SIHREN *application generated with* GENGHIS

5.6.1.2. *The* MOVISS *application*

The MOVISS project [BEC 10] focuses on the implementation of methods and tools to assess a town's social vulnerability by taking into account the mobility of individuals during the daytime. The goal was to have a tool allowing us to build scenarios to diminish social vulnerability by causing the different variables that have an influence on the social vulnerability index (SVI) [BEC 09] to fluctuate.

In MOVISS [BEC 10], the data used are relatively homogeneous and structured: it is attributary data defined along a unique geographic grid (the neighborhood) and a temporal granularity corresponding to periods: morning, noon, and evening. For each geographic entity and each temporal period, a vulnerability index is calculated based on different indicators, such as knowledge, perception, and information levels. All these levels are weighted. The indicators are themselves defined through statistic processing carried out on data from sociological surveys. The variation in the value of these indicators, as well as the variation in weight, influences the SVI value and thus the spatiotemporal distribution of social vulnerability.

After defining the adequate data model, we have designed through GENGHIS an application allowing us to visualize SVI and its linked indicators spatiotemporally. The spatiotemporal analysis carried out in MOVISS requires us to have an interface allowing for the visualization of data depending on the three study periods (morning, noon, and evening) simultaneously, to identify spatial differentiations for each studied period as well as temporal differentiations for each geographic entity. This entails a double-entry cartographic reading relying on the principle of map collection. Thus the geovisualization application developed in GENGHIS is structured into four frames (see Figure 5.15):

– an information frame displaying for each geographic entity, in a table, the value of the vulnerability indexes and the associated indicators depending on the three periods studied (morning, noon, and evening);

– three spatial frames matching each of the temporal periods that display simultaneously as a choroplete maps, the spatial distribution of the Grenoble neighborhoods' SVI or associated indicators. These frames have the viewing functionalities offered by GENGHIS (zoom, *pan*, layer management, etc.); they are interconnected and synchronized among themselves, allowing for dynamic and interactive viewing of the data.

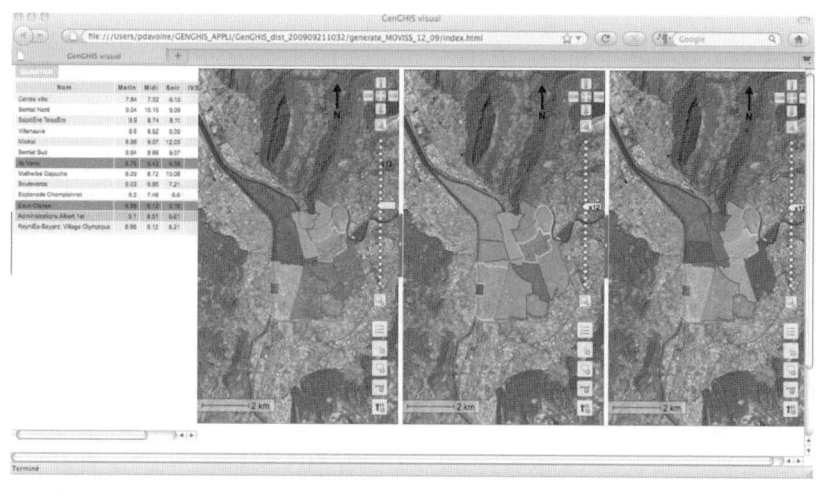

Figure 5.15. MOVISS *application interface generated by* GENGHIS

Given the type of temporality characterizing MOVISS's data, using the temporal frame as it is initially suggested by the GENGHIS environment is not necessary.

As for building vulnerability diminishing scenarios, we have integrated a new functionality allowing the user to interactively modify the indicator values, whether they are the

SVI itself or indicators influencing the SVI (see Figure 5.16). Owing to the MOVISS application, we have been able to generate a tool that genuinely helps decision making.

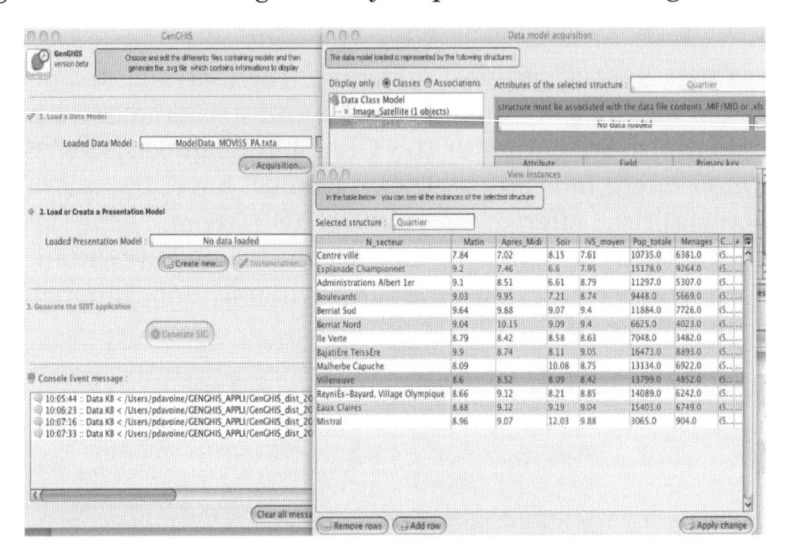

Figure 5.16. *Typing interface allowing us to run vulnerability scenarios for the* MOVISS *application*

5.6.2. *User community*

Originally stemming from issues in managing data linked to natural risks and more specifically data linked to past catastrophic event characterization, the GENGHIS platform appeals to users with a heterogeneous spatiotemporal or spatio-historical geographical information that also has a multimedia dimension. This is notably the case for geoscience communities (seismologists, geologists, hydrologists, etc.) that use a great amount of multidimensional environmental data as well as all the communities managing georeferenced historical information (geographers, historians, etc.).

GENGHIS is suited for the development of applications allowing us to take into account orthogonally spatial, temporal, documentary, and informational dimensions. It appears that this approach is useful to understand and analyze environmental phenomena which are often based on the use of heterogeneous spatiotemporal or spatiohistorical information integrating a multimedia dimension. GENGHIS can thus appeal to the geoscience community (seismologists, geologists, hydrologists, etc.) that uses a lot of multidimensional environmental data, as well as for geographers or historians managing georeferenced historical information.

GENGHIS is developed by computer specialists for non-computer specialists. The final users are currently mostly researchers, but in the long run this tool must also be aimed at managers. However, its use requires sound knowledge of the modeling tool and processes, especially in AROM-ST. Indeed, the design of the data model is not directly done by GENGHIS and knowledge of the modeling process in AROM and AROM-ST is required.

GENGHIS's development was mainly carried out by engineering students at the CNAM (French National Conservatory of Arts and Crafts) and by computer science students at master level. The approach chosen by the GENGHIS environment also raises graphical semiology and cartographic representation issues, so the duty of specifying these aspects, and notably spatiotemporally visualizing information, is given over to geographers.

5.7. Conclusion and perspectives

Recent technological breakthroughs have considerably increased the need for geovisualization tools that allow us to use heterogeneous and multimedia data referenced both in time and space. The development of a computing environment such as GENGHIS falls within the parameters of this need to help create geovisualization applications that are suited to users' needs and manipulated data characteristics. GENGHIS was presented in 2009 at the International Festival of Geography of Saint-Dié des Vosges [DAV 09]. Members of the geovisualization competition acknowledged that it was a very promising tool opening many application perspectives in the field of the environment as well as that of territory management. Simultaneous visualization of different dimensions of the manipulated information, interactivity and dynamic synchronization of the frames, as well as the modeling approach underpinning the GENGHIS application, enables us to build a geovisualization interface that responds to the needs of a user community for which classical GIS are too complex and not always well suited. However, this tool does have limits and still requires many functionalities to be developed. It is therefore the focus of many research and development perspectives, in the field of modeling and spatiotemporal knowledge representation as well as geovisualization. Among these we can mention a few: integrating the diversity of temporalities, the notion of quality, and even of uncertainty in geographical data; adapting graphical semiology and cartography to the user diversity; integrating rules for cartographic design; integrating modules to help design geovisualization interfaces.

Within the current ANR projects, other GENGHIS applications are considered (the URBASIS project of the RiskNat ANR or the Biblindex project studying spatiotemporal dissemination of biblical texts), opening up new opportunities to develop innovative functionalities that can be pooled.

5.8. Acknowledgments

The development of GENGHIS was carried out thanks to the funding of the regional council of Isère (through the Grenoble center for natural risks), the Rhône-Alpes region, the Joseph Fourier university, and the Department for Environment, Energy, Sustainable Development and the Sea. These development were carried out thanks to different projects funded by the ANR (Urbasis project of the RiskNat ANR, Biblindex project of the ANR Programme Blanc) and by the "Environment" cluster of the Rhône-Alpes region (PercuRisk project).

We would also like to thank researchers and partners of all these projects who took part in GENGHIS's specifications (PACTE-Territoires laboratory, LIRIGM, LGIT, LEPPI, CEMAGREF, PGRN, Acthys Diffusion, the city of Grenoble, and the urban community of Grenoble-Alpes Métropole), as well as the students, interns, and engineers who contributed to the development of the different versions of the software.

5.9. Bibliography

[ARN 09] ARNAUD A., Valorisation de l'information dédiée aux événements de territoires à risque. Une application sur la couronne grenobloise, PhD thesis, University Joseph Fourier, Grenoble, 2009.

[BEC 09] BECK E., ANDRÉ-POYAUD I., CHARDONNEL S., DAVOINE P.-A., LUTOFF C., "Spatio-temporal variations in the vulnerability to earthquakes in Grenoble (France)", *16th European Colloquium on Quantitative and Theoretical Geography*, Maynooth, Ireland, 2009.

[BEC 10] BECK E., ANDRÉ-POYAUD I., CHARDONNEL S., DAVOINE P.-A., LUTOFF C., MOVISS: Méthodes et Outils pour l'évaluation de la VulnérabIlité Sociale aux Séismes, final report, programme du pôle grenoblois des risques naturels, 2010.

[CAR 09] CARTIER S., BECK E., BOUDIS M., CORNOU C., DAVOINE P.-A., GUEGUEN P., SAILLARD Y., SIRSEG: Simulation du risque sismique et de ses enjeux à Grenoble, rapport final, programme de recherche "Risque – Décision – Territoire" du MEEDDM, 2009.

[CLU 98] CLUET S., "Designing OQL: allowing objects to be queried", *Information Systems*, vol. 23, no. 5, pp. 279–305, 1998.

[DAV 03] DAVOINE P.-A., BRUNET R., "SIDIRA: un système d'information basé sur le Web dédié à la consultation des événements avalanches: application à la commune de Vallorcine", *SIRNAT 2003*, Orléans, France, January 2003.

[DAV 04] DAVOINE P.-A., MARTIN H., CŒUR D., "Historical flood data base linked to a web-based interface", *Sytematic, Palaeoflood and Historical Data for the Improvement of Flood Risk Estimation (SPHERE), Methodological Guidelines*, CSIC Madrid, Spain, pp. 95–101, 2004.

[DAV 06] DAVOINE P.-A., MOISUC B., GENSEL J., MARTIN H., "SIHREN: conception de systèmes d'information spatio-temporelle dédiés aux risques naturels", *Revue Internationale de Géomatique*, vol. 16, nos. 3–4, pp. 377–394, 2006.

[DAV 09] DAVOINE P.-A., MOISUC B., GENSEL J., ARNAUD A., "GenGHIS: environnement pour le développement d'applications de géovisualisation à données géo-référencées multidimensionnelles", *Festival International de Géographie de Saint-Dié des Vosges, Salon de la Géomatique, Concours de Géovisualisation*, October 2009.

[GAY 09] GAYET L., Réingénierie d'un générateur d'applications de système d'information spatio-temporelle: GenGHIS, mémoire d'ingénieur, CNAM Grenoble, 2009.

[MOI 07] MOISUC B., Conception et mise en œuvre de systèmes d'informations spatio-temporels adaptatifs: le framework ACTIS, PhD thesis, University Joseph Fourier, Grenoble, 2007.

[OGC 02] Open Geospatial Consortium, Styled Layer Descriptor Implementation Specification, version 1.0.0, 2002.

Chapter 6

GEOLIS: a Logical Information System to Organize and Search Geo-Located Data

6.1. Introduction

GEOLIS is a geo-located data exploration tool. It concerns files of the GML format and presents itself as a Web interface combining a query zone, a dynamic map displaying the query results, and a dynamic index reflecting the result distribution. GEOLIS's main specificity is to guide the user step-by-step in building complex queries while guaranteeing that they will not lead to empty results. This navigation allows both targeted and exploratory searches. GEOLIS is mainly meant for final users, but also offers several expansion points so as to specialize it for different applications.

Chapter written by Olivier BEDEL, Sébastien FERRÉ and Olivier RIDOUX.

6.2. Background history

GEOLIS was developed by Olivier Bedel during his doctorate with the Logical Information Systems (LIS) team at the IRISA[1] between 2005 and 2008. Its development was then taken up by Pierre Allard. GEOLIS follows on from the development of LIS and coexists with other implementations (CAMELIS, LISFS, ABILIS) which share some components. The original idea behind LIS was suggested by Olivier Ridoux who found that hierarchical data organizations, in general, and hierarchical file systems, in particular, were unsatisfactory. He was not satisfied with databases either due to their lack of flexibility, lack of integration in a system and with other applications, or lack of navigation to help non-expert users. His first idea was to combine query expressiveness, by using logics, and navigation ease, by suggesting query increments. The successive doctoral dissertations of Sébastien Ferré [FER 04] and Yoann Padioleau [PAD 05a] laid down the theoretical and practical foundations for LIS and developed the first prototypes. Contacts with geographers from the RESO laboratory at the University of Rennes 2 allowed us to establish the relevancy of LIS for geographic data exploration. In 2005, the LIS team partnered with Erwan Quesseveur and François le Prince to obtain a PhD grant from the region of Brittany, which funded Olivier Bedel's doctorate [BED 09]. This chapter gives a summary of the main contributions of this dissertation by presenting the major functionalities and the architecture of GEOLIS. A few "user" cases are given to illustrate the chapter and GEOLIS's extension points are highlighted.

---

[1] The IRISA is a mixed research unit (UMR 6074) partnered with institutions such as the CNRS, the University of Rennes 1, and the ENS Cachan.

6.3. Main functionalities and use cases

GEOLIS's main function is to explore a set of geographic data. This notion of exploration covers a large scope going from a direct search by formulating a query to simple forms of data mining. To this end, GEOLIS is based on LIS [FER 04, FER 09]. This involves a certain data structure which is different from the usual layout structure found in geographical information systems (GIS). The geographic entities belonging to the different layers are gathered in a single set of objects. Each object is linked to properties describing the geometry and thematic attributes of this object. Object descriptions do not have to follow a common pattern, which makes processing heterogeneous data easier. The notion of a layer becomes virtual. A layer is defined as the set of geographic entities verifying a specific combination of properties: for example, "the grain parcels with an area over one hectare". Thus, the original layers can be found and a multitude of other layers can be defined. The different layers obviously do not have to be unconnected. Characterizing a layer is a query whose answers are the entities making up the layer.

A core element of this approach is the language used to represent object properties, as well as links which can exist between objects. We have decided to use logical languages in LIS, since logic adds reasoning mechanisms to property representation allowing us to infer implicit properties based on explicit properties. For example, if we have the `area:2a` property for a parcel-object, logic allows us to infer an infinity of new properties such as `area:>=1ha` or `area:[1ha,5ha]`. Indeed, inferred properties are more general than the explicit property. We can define the type of logics expected by LIS as follows:

DEFINITION 6.1.– *A LIS logic is a partially ordered set* $\mathcal{L} = (L, \sqsubseteq)$ *in which:*

– L *is a formula language used to describe objects. L defines all the expressions allowing us to build a formula, for example* `area:>=1ha`.

– *The partial relation* $\sqsubseteq$ *is called subsumption relation. It matches a generalization/specialization relation between formulas. For example,* `area:2ha` $\sqsubseteq$ `area:>=1ha`.

These logics are equipped with semantics which provide meaning to the formulas and a subsumption specification [BED 09, FER 04]. We only give the logic semantics presented in this chapter in an informal manner since they are more about the design of logics than about their use.

A GEOLIS database is called a "logical context". A logical context is the data structure on which LIS are based. It is made up of two subcontexts: one to describe objects and the other to describe relationships between objects. Each subcontext has its own logic since object and relationship descriptions are generally different. Section 6.3.2 describes geometry logics for objects, and distance and topology logics for relationships between objects.

DEFINITION 6.2.– *A logical context is a* $K = (K_1, K_2)$ *pair, in which:*

– $K_1 = (\mathcal{O}, \mathcal{L}_1, d_1)$ *is the object context, with* $\mathcal{O}$ *all the objects,* $\mathcal{L}_1 = (L_1, \sqsubseteq_1)$ *is the logic used to describe the objects, and* $d_1 \colon \mathcal{O} \to L_1$ *is a function mapping each object to its logical description.*

– $K_2 = (\mathcal{R}, \mathcal{L}_2, d_2)$ *is the relationship context, with* $\mathcal{R} \subseteq \mathcal{O} \times \mathcal{O}$ *a set of* (o, o') *couple of objects, ties (oriented) between objects,* $\mathcal{L}_2 = (L_2, \sqsubseteq_2)$ *is the logic used to describe each couple of objects, and* $d_2 \colon \mathcal{R} \to L_2$ *is a function mapping each pair of objects which is logical description.*

A logical context can be seen as a graph whose nodes are the objects and whose arcs are the links between the objects. The nodes are labeled with $\mathcal{L}_1$ logic formulas, called "properties", while the arcs are labeled with $\mathcal{L}_2$ logic formulas, called "relationships". One of the GEOLIS's advantages is to be able to choose different logics depending on each dataset or application. This choice of logics replaces the diagram conception in the relational database. To fix ideas, the current version of GEOLIS allows us to describe objects and links between objects as sets of elementary properties which can either be simple (boolean) attributes or valued attributes. The values belong to different value fields, such as numbers and intervals, character chains, and geometries (e.g. points, lines, and polygons). Concrete formula examples are introduced in section 6.3.1 in an intuitive mode, and spatial logics are more closely defined in section 6.3.2. Table 6.1 recaps how GEOLIS's data model matches the layer structure.

GEOLIS	Layer structure
Object	Geographic entity
Logical language	Layer description diagram
Logical property	Thematic attribute, spatial description, metadata
Logical relation	Link between entities
Query	Virtual layer
Logical context	Set of data, set of layers

Table 6.1. GEOLIS *data model vs. layer structure*

To illustrate GEOLIS's major functionalities and use cases, we have chosen a set of real but relatively simple data describing an island off the coast of Britanny, Milliau[2]. The island of Milliau is off the coastal town of Trebeurden in the

---

[2] The authors would like to thank Erwan Quesseveur of the UMR ESO at the University of Rennes 2 for creating this dataset and making it available.

Côtes d'Armor region (48°46'9"N, 3°35'51"O). The original dataset is made up of 11 thematic layers representing the limits of the island, the buildings present, the sightseeing points of interest, the paths, and the vegetation. All in all, 49 geographic objects are described with three attributes each on average. Spatial relations were calculated for each couple of objects of the dataset. When integrating the dataset into GEOLIS, all the layers were merged into a single set of objects. Each object was described by a property characterizing its type (such as building, road, and rock), determined from the layer it belonged to. For example, here is the description for objects number 1 and number 2 as well as their link:

– $d_1(1)$ = {ruin, nature:"Aristide Briand house", description:"360 degrees Point of View", disabled_access:"Yes", area:282.36m2, length:79.36m, shape:{Polygon}} ;

– $d_1(2)$ = {cultural heritage, nature:"wash house", disabled_access:"No", shape:{Point, Convex, #0-Edge}} ;

– $d_2(1,2)$ = {distance:290.8m}.

6.3.1. *Geographical data visualization and exploration*

In most GIS, an information search is carried out by starting from the primitive layers stored on the disc and applying operations such as filters or transformations. These operations create new layers on which we can then apply operations again. In GEOLIS, the layers are virtual and characterized by a query. Section 6.3.1.1 defines the query language. Each query is the intention of a concept for which we are seeking the extension, that is the set of entities of the layer characterized by the query. Section 6.3.1.2 describes the presentation of a virtual layer for users. These presentations obviously include a cartographic view as well as an index of the layer entities properties. Section 6.3.1.3 shows how these views can be used by users to build and explore complex virtual layers by simple navigation. The navigation links define query transformations which play a similar role to operations as on GIS layers.

6.3.1.1. *Virtual layers: queries and extensions*

Given a logical context, each logical property defines a virtual layer: all the entities have this property. In a similar way, each logical relation defines two virtual layers: the domain and the co-domain of the relation. These elementary virtual layers can be combined by set-theoretical operations and turn into queries which themselves combine properties and relations.

DEFINITION 6.3.– *Let $K = (K_1, K_2)$ be a logical context. If $p \in L_1$ is a logical property, $r \in L_2$ is a logical relation, and q_1, q_2 are queries, then we can make the following queries (followed by their information signification):*

ALL	*any object*
p	*an object described by the property p*
$r \rightarrow q_1$	*the origin of a link of the relation r whose destination is q_1*
$r \leftarrow q_1$	*the destination of a link of the relation r whose origin is q_1*
NOT q_1	*not q_1*
q_1 AND q_2	*q_1 and q_2*
q_1 OR q_2	*q_1 or q_2*

All these queries make up the L language.

The formal signification of the virtual layer characterized by a query can be defined by all the geographic entities belonging to this layer, which is called an extension. In terms of information searching, this extension matches all the answers to the query.

DEFINITION 6.4.– *Let $K = (K_1, K_2)$ be the logical context. If $p \in L_1$ is a logical property, $r \in L_2$ is a logical relation, and*

q_1, q_2 are queries, then a query's extension in K is recursively defined by:

$$\begin{aligned}
ext(\text{ALL}) &= \mathcal{O} \\
ext(p) &= \{o \in \mathcal{O} \mid d_1(o) \sqsubseteq_1 p\} \\
ext(r \rightarrow q_1) &= \{o \in \mathcal{O} \mid \exists o' \in \mathcal{O} : d_2((o, o')) \sqsubseteq_2 r,\ o' \in ext(q_1)\} \\
ext(r \leftarrow q_1) &= \{o' \in \mathcal{O} \mid \exists o \in \mathcal{O} : d_2((o, o')) \sqsubseteq_2 r,\ o \in ext(q_1)\} \\
ext(\text{NOT } q_1) &= \mathcal{O} \setminus ext(q_1) \\
ext(q_1 \text{ AND } q_2) &= ext(q_1) \cap ext(q_2) \\
ext(q_1 \text{ OR } q_2) &= ext(q_1) \cup ext(q_2)
\end{aligned}$$

We note that the content of a virtual layer is always a set of geographical entities and that belonging or not to a layer's entity only depends on its properties, its linked entities, and whether or not these entities belong to other layers. For example, the query:

```
cultural heritage AND distance:([0.,49.99])
 -> (building AND disabled_access:"Yes")
```

defines the layer with the points of interest which are less than 50 m from a building with disabled access.

6.3.1.2. *Visualizing a virtual layer: map and navigation index*

A virtual layer's layout in GEOLIS is made up of three views (see Figure 6.1 showing the virtual layer of objects offering a disabled access):

– (above) the query characterizing the layer;

– (in the middle) the map displaying the layer entities (the query extension) depending on their position and geometry;

– (left tree structure) the index which is an inventory of the layer entities according to their different properties.

The query and the map are classical elements in GIS. The index is the part which is specific to GEOLIS and LIS in general. It can be seen as an enriched and dynamic type of legend: enriched since it shows all the properties and relations defined on objects and dynamic since its content is calculated according to the virtual layer's content. It is made up of both alphanumerical properties (such as `cultural heritage, nature:"covered lane", area:([0.,9.99])`) and geometric properties (such as two rectangles on the map outlining the areas of interest). The first are organized along a tree structure while the latter are directly projected onto the map.

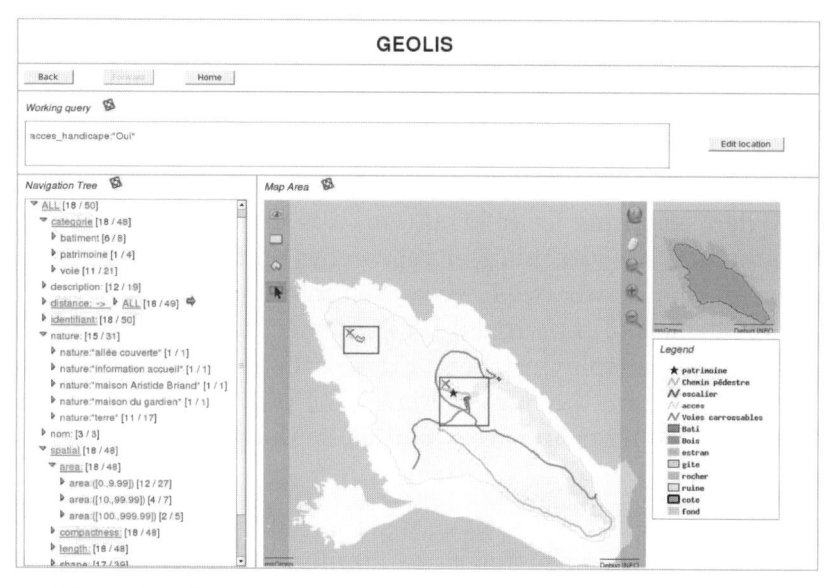

Figure 6.1. *View organization in the* GEOLIS *interface. For a color version of this figure, see www.iste.co.uk / Bucher / innovgis.zip*

6.3.1.2.1. Map view

The cartographic view is built up from the geometric description of the entities in the layer. It represents these entities in a geographical context defined by:

– A system of coordinates specific to the cartographic view and determined for the dataset.

– A backdrop defining the cartographic frame in which the objects of the layer are placed. The aim of the backdrop is defined by a general geographic frame. It can, for example, be a satellite image and an aerial photograph.

– A legend qualifying symbology, which is the graphical representation associated with each category of objects represented on the map.

Cartographic view is not a static representation. The user has standard cartographic navigation functionalities at his/her disposal (see Figure 6.1): zoom in, zoom out, zoom box, *pan*. A keymap reminds the user of the current zoom level on a wider geographical zone, traditionally corresponding to the initial zoom.

When searching for information, it is important to know to what extent the virtual layer represents all the entities in the dataset. That is why we have chosen to put the entities which do not belong to the virtual layer in the background instead of simply not displaying them. Finally, the rectangular frames which appear on the map in Figure 6.1 play the same role as the index elements and are explained in the following sections.

6.3.1.2.2. Index view

A virtual layer's index is similar to a book's index where the terms are queries and the pages are layer entities. Each index query is called "increment" and characterized by another virtual layer having a non-empty intersection with the indexed layer. Figure 6.1 shows the `disabled_access:"Yes"` answer and more specifically the indexation of objects making up this answer. In this index, we see the `building, cultural heritage,` and `road` categories appear. This means that these different categories have objects which are in the answer. However, not all of the original set categories appear (such as `forest` and `rock`) since in those categories there is no object answering the query.

DEFINITION 6.5.– *Let $K = (K_1, K_2)$ be a context and q a query. A query x is an increment of the q layer iff $ext(q) \cap ext(x) \neq \emptyset$.*

The first difference with a book index is that the GEOLIS index is not calculated once and for all for the dataset, but calculated layer-by-layer and then updated to precisely reflect the content of the indexed layer. A second difference is that instead of listing, for each increment, the common entities with the indexed layer q, an increment is decorated with: the ($n_{q \cap x} = \|ext(q) \cap ext(x)\|$) number of entities common to the two layers, the ($n_x = \|ext(x)\|$) number of entities in the increment layer. We easily get the ratios $r_{x/q} = n_{q \cap x}/n_{q \cap ALL}$ and $r_{q/x} = n_{q \cap x}/n_x$ which allow us to establish the relation between two layers. If $r_{x/q} = 1$, then layer q is included in the layer x (e.g. x = distance:([0.,49.99]) -> road: all disabled accesses are less than 50 m from a road). Inversely, if $r_{q/x} = 1$, then the layer x is included in the layer q (e.g. x = gite: all the gîtes have a disabled access). If both ratios are equal to 1, then both layers are equivalent and they contain the same entities.

To limit the index calculation and representation complexity, the increments are limited to queries under the p, r -> p and r <- p shape, where p is either a logical property or ALL, and r is a logical relation. This is sufficient to visualize the properties of the layer's objects as well as those of the objects direction linked to those on the layer. We will show in section 6.3.1.3 that these increments are also sufficient to build queries with arbitrary embedding depth.

The index increments are not independent and can be classified according to a generalization order. We can therefore define a subsumption relation $\sqsubseteq$ between queries, relying on subsumption relations between properties on the one hand and between relations on the other (see Definition 6.2).

For example, we have the following relation:

```
distance:39.2 m -> gite ⊑ distance:(<=50 m)
                -> building
```

Indeed, to be at 39.2 m of a gîte implies being less than 50 m away from a building. This definition wishes to be simple and does not cover all cases. For example, we do not have `gite ⊑ (building OR cultural heritage)` whereas all gîtes are buildings. But it is sufficient to classify the index as well as to define the navigation modes (section 6.3.1.3).

The index is thus in a tree shape with each node representing an increment and its child node–parent node relation matches the subsumption relation. Figure 6.2 shows an example of this tree classification. Each increment is decorated by the ratio $r_{q/x}$. When the ratio equals 1, the increment is underlined to indicate that all objects in the layer are covered by this increment. The user can unfold or fold each node of the tree to display or hide more specific increments. A relational node, i.e. a node with a relation $r \rightarrow p$ (respectively, $r \leftarrow p$), can be unfolded in two ways: to specify the type of relation r or to specify the range (respectively, domain) p of the relation. This node thus has two folding/unfolding icons. For instance, in Figure 6.2, the relational node `distance -> ALL` characterizes all the objects for which a distance relation with another object has been filled out. On the one hand, this node is unfolded versus the distance relation: distance intervals are suggested, such as `distance:([0.,49.99])` or `distance:([50.0,99.9])`. On the other hand, this node is also unfolded versus the property ALL for which more specific properties are offered (such as `distance:->category`, `distance:->description:` or yet `distance:->equipment:`).

Increments of a geometrical nature are directly represented on the map with the same color symbology and the same folding/unfolding rules. Figure 6.1 shows two of these increments on the map (dark gray rectangles).

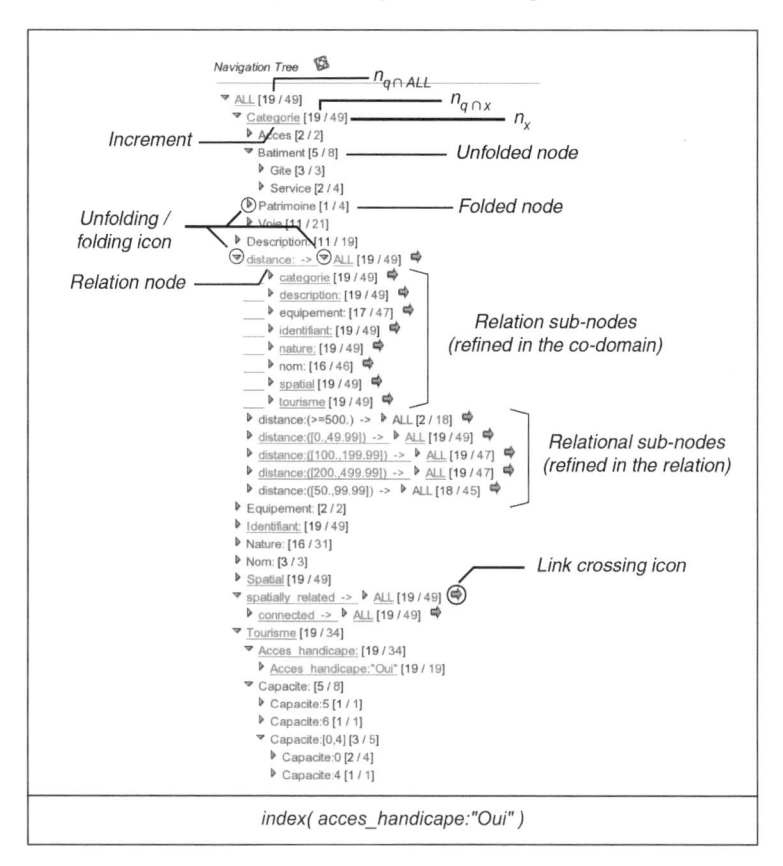

Figure 6.2. *Index view of the* disabled_access:"Yes" *query. Underlined are a relation node such as* distance -> ALL *replaces the parent node relation and thus avoids visually overloading the navigation index. For a color version of this figure, see www.iste.co.uk/Bucher/innovgis.zip*

6.3.1.3. *Building and transforming virtual layers: navigation links*

The previous section showed how the content of a virtual layer can be visualized thanks to the map and the index, but not how users can reach this layer. The two ways offered by GEOLIS are querying and navigating. Querying is the most direct way of reaching a layer. The user types in a query characterizing the desired layer and the two map and index views are recalculated. Geometrical constraints such as "within such a polygonal region" can be directly defined on the map by drawing a polygon. It is indeed very difficult for a user to manually type a polygonal description. However, querying has three growing difficulties for users:

– A user who does not know GEOLIS does not know the query language syntax.

– A user who knows GEOLIS but does not know the dataset knows the query language but does not know what the available properties and relations are.

– A user who knows GEOLIS and the dataset can ask syntactically accurate queries with no guaranteed result since he/she does not know all the valid property and relation combinations (if he/she did, he/she would not need an information system).

Therefore, GEOLIS offers, from any layer, a set of navigation links toward other layers. Each navigation link defines a query transformation characterizing the current layer, and producing a new query and thus a new layer. The user is thus guided into building the desired layer. Actually, as we will see in section 6.3.3, rather than a way to reach a layer, navigation should be seen as a way to explore a dataset. Each navigation stage brings its own set of information independently from the goal pursued and there can even be no precise goal. GEOLIS navigation has the following advantages (demonstrated in section 6.3.1.3.4):

– It guarantees never to reach empty layers.

– It allows us to make any query with no negation, no disjunction, and whose properties and relations appear in the index.

A user can thus create complex queries simply by clicking on navigation links. Even if he/she does not know anything about the dataset, he/she will never end up in a dead end. Indeed, GEOLIS relies on user's ability to understand the signification of queries and navigation links which is greater than their ability to enunciate these queries. If a query is not entirely accessible through navigation, it can be edited and we can apply manual transformation on the layer. For example, the query road AND NOT connected -> access ("a road unconnected to an access") can be reached by inserting NOT in the road AND connected -> access query, which is accessible through navigation. Thus, navigation and query can be mixed into one single search.

We will later detail all the different types of navigation links, which match different types of query transformations, and thus of layer transformations. In the interface, the navigation links are active elements in three views: index increments (in the tree or on the map), and query parts. To define the query transformations, it is sometimes useful to see the query in its normal conjunctive shape, i.e. as a set of queries connected by AND. For example, the query with a negation above can also be presented as { road, NOT connected -> access }.

6.3.1.3.1. Refining and expanding

"Refining" the current query q with an increment x causes the current layer to be restricted to the objects present both in q and in x. Formally, this is equivalent to replacing the elements of q which subsume x with x:

$$refine(x) = q \rightsquigarrow (q \setminus \{y \in q \mid x \sqsubseteq y\}) \cup \{x\}$$

On the contrary, "expanding" the query $q = q_1$ AND y with an increment x which is more general than y ($y \sqsubseteq x$) is equivalent to expanding the current layer to the objects present in q_1 and x: the property of y was generalized into x. If $y = x$, then q is generalized by taking y away from q. Formally:

$$widen(x) = q \leadsto \begin{cases} q \setminus \{x\} & \text{if } x \in q \\ (q \setminus \{y \in q \mid y \sqsubseteq x\}) \cup \{x\} & \text{if not} \end{cases}$$

At the level of the GEOLIS interface, refining or expanding is caused by selecting an increment. If the increment is not-underlined (see Figure 6.2), then it is not shared by all the objects in the current layer, and thus causes refining to take place. If the increment is underlined (see Figure 6.2), then it is shared among all the objects of the current layer, and it causes an expansion to take place. The process is the same whether it is a thematic increment that appears in the tree structure or a geometric increment drawn on the cartographic view. Refining with a geometric increment allows us to diminish the current selection to objects in the geographic region outlined by this navigation link. In this sense, this type of refining is called "logical zoom", to separate it from cartographic zoom.

6.3.1.3.2. Link crossing

Relation navigation corresponds to a change in points of view on the dataset. A link shaped as $r \rightarrow p$ can be used to cross the relation r from the current query q. This is equivalent to selecting among the images with r objects of the current layer which are also described with the property p. In a symmetrical way, relations can be crossed in the opposite direction by using increments shaped as $r \leftarrow p$. In a more formal manner:

$$trav(r \rightarrow p) = q \leadsto p \text{ AND } r \leftarrow q$$
$$trav(r \leftarrow p) = q \leadsto p \text{ AND } r \rightarrow q$$

Crossing a relation is caused by activating the crossing icon matching each relational increment in the index (see Figure 6.2).

6.3.1.3.3. Query reversing

When a query is shaped as $q = q_1$ AND $r \rightarrow q_2$, the current layer matches all the objects in the layer q_1 which are each in a relation r with at least one object in layer q_2. The query q corresponds to the vision centered on q_1 of the relation r. Reversing this query means pivoting around the relation r to center the new query around q_2. More formally[3]:

$$rev(r \rightarrow q_2) = q \rightsquigarrow q_2 \text{ AND } r \leftarrow (q \setminus \{r \rightarrow q_2\})$$
$$rev(r \leftarrow q_2) = q \rightsquigarrow q_2 \text{ AND } r \rightarrow (q \setminus \{r \leftarrow q_2\})$$

Reversing the query q on a relational branch $r \rightarrow q_2$ happens in the query view. Each relational branch of the current query susceptible to be reversed is underlined, like a hyperlink (see Figure 6.3). These branches are active, meaning these can be selected by clicking, which reverses the query.

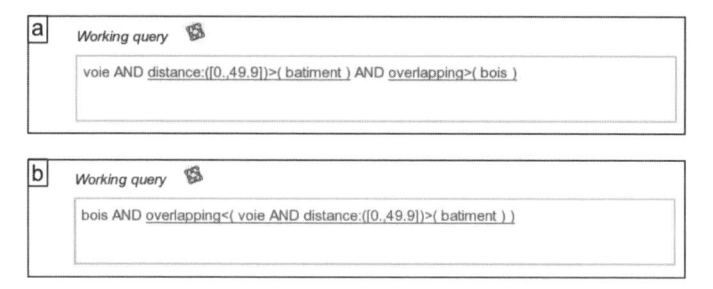

[a] Working query
voie AND distance:([0.,49.9])>(batiment) AND overlapping>(bois)

[b] Working query
bois AND overlapping<(voie AND distance:([0.,49.9])>(batiment))

Figure 6.3. *Query reversing interface*

6.3.1.3.4. Accuracy and completeness of navigation links

The navigation links defined above are accurate inasmuch as, starting in a non-empty layer, they never lead to an empty layer. Navigation links are also compete to build any query

---

3 In these expressions, q is considered in its normal conjunctive shape and $q \setminus \{r \rightarrow q_2\}$ refers to q without its subquery $\{r \rightarrow q_2\}$.

without negation or disjunction using only properties and relations present in the index. In other words, any query verifying the previous criteria is reachable by navigation. Obviously, this is only true for queries that define a non-empty layer.

6.3.2. *Representation of geographical data and spatial reasoning*

The logics used by GEOLIS are not fixed; they can be extended with components specific to such and such application. These logical components belong to a common pattern which ensures a certain compatibility with GIS data description patterns while allowing greater flexibility. Each object of a logical context corresponds to a geographical entity whose logical description is a set of properties and logical relations. The properties stem from the original layer, the entity's geometry and the thematic properties, whereas the relations stem from entity matching tables. Each property or relation is represented as a simple attribute or a valued attribute. Some attributes are universal, such as geometry or distance, while most are specific to each application and stem from the layer diagrams (for instance, category, nature, and capacity). Some attributes can be derived from other attribute calculations, such as an entity's area or the distance between entities from geometry. Each attribute has its own value domain, for example, nature takes character chains, capacity uses integers. Each value domain can be modeled as a logic and implemented as a logical component which can be integrated into GEOLIS. The rest of this section defines the logics for value domains with a spatial aspect: geometry, area, perimeter, shape, distance, and topological relation. Other logics exist for integers, character chains, and dates [BED 09].

6.3.2.1. *Representing spatial properties*

The main spatial property is geometry, which defines the position, nature (such as point, line, and region), and shape of an entity. The other spatial properties are actually derived from geometry. They allow us to express entity properties which are unvarying in their position and orientation: area (for regions), perimeter/length (for regions/lines), and shape (e.g. dimensions, convexity, and number of sides).

6.3.2.1.1. Geometry

The $\mathcal{L}^{G\subseteq}$ logic is a logic of geometry shapes in a two-dimensional space. It matches the reserved attribute geometry. The $L^{G\subseteq}$ formulas are WKT language expressions (the acronym for well-known text) [HER 06] which offer a textual representation of the geometry model suggested by OGC. Here are some examples of geometries described in WKT format:

POINT(6 10)
  A point is a coordinate with two components.
LINESTRING(3 4,10 50,20 25)
  A broken line is a sequence of coordinates.
POLYGON((1 1,5 1,5 5,1 5,1 1),
  (2 2, 3 2, 3 3, 2 3,2 2))
  A polygon is a series of rings (closed broken lines); the first ring refers to the outside border of the polygon, the following its inside border. It is thus possible to define polygons with holes.

The subsumption relationship $\sqsubseteq^{G\subseteq}$ orders WKT expressions according to the inclusion relation between geometries. A g_1 geometry is included in a g_2 geometry if all points of g_1 are also points of g_2. This order relation on geometries is illustrated in Figure 6.4. Each WKT expression can be used as a spatial inclusion pattern allowing us, for example, to limit the information search to a zone of interest (the map in Figure 6.1 has two).

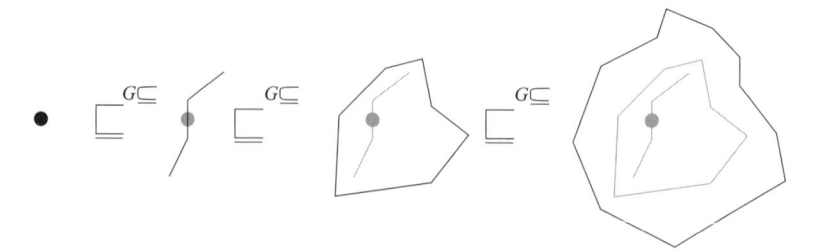

Figure 6.4. *Order relation on the geometric representations. The grayed out geometries show the included geometries*

6.3.2.1.2. Derived spatial properties

The three logics $\mathcal{L}^{Area}$, $\mathcal{L}^{Length}$, and $\mathcal{L}^{Shape}$ enable us to represent, respectively, the area, length, and shape of a geographic entity. These three value domains are matched to the reserved attributes area, length, and shape. Deriving these three attributes from the geometry of an entity relies on three functions to calculate its value. The area and the length are naturally real numbers (matching a certain unit), while the shape is a set of descriptors among:

– *empty*, *point*, *line*, and *polygon* characterizing the size of the geometry;

– *convex* indicating that no straight line is continuous, the sides of the geometry cross it, and *concave* indicating the opposite;

– *equilateral* indicating that all the sides of the geometry are equal;

– *rightAngle* meaning having at least one right angle;

– *regular* implying *equilateral* with all the angles also being equal;

– *n-edge* specifying the number n of sides in the geometry.

We can directly express a value with a geometric description from which this value will be calculated. In a search for information, this allows us to search for all the objects with the same area as a given geometry which can be directly drawn on the map. Different units are available for numerical values and conversions are automatically carried out when necessary. Moreover, to increase the expressiveness of queries, we can use value intervals (including as geometries) for area and length, and descriptor subsets for shape. For example, area:(<= 100 m2) selects the entities whose area is under 100 m^2 and shape:{#4-Edge, Regular} selects the square-shaped entities. These patterns (intervals and subsets) structure the subsumption relation. An interval is subsumed by another interval if it is included in this interval. A set of shape descriptors is subsumed by another set of descriptors if it contains this other set. These subsumption relations are illustrated in Figure 6.5.

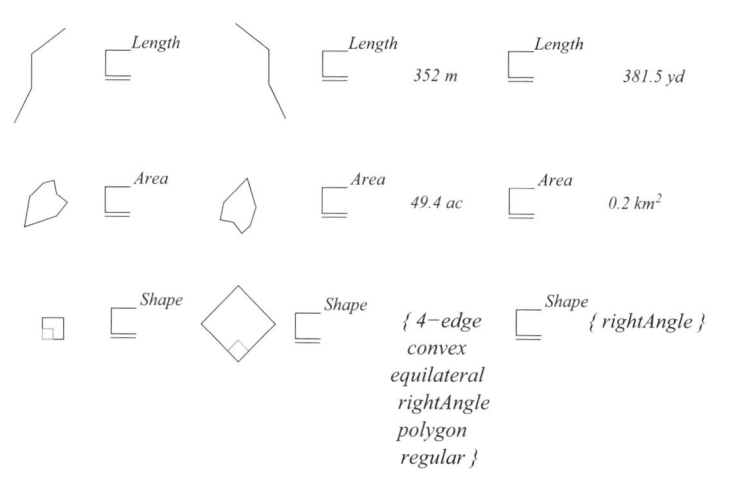

Figure 6.5. *Subsumption relations between the area, length, and shape logic formulas. The* WKT *expressions are graphically represented*

6.3.2.2. *Representing spatial relations*

Among the spatial relations, some, like distance and direction, are often qualified by a measurement, and are then called quantitative relations. Other topological relations such as adjacency, overlapping, and inclusions describe an abstract spatial organization often close to our cognitive perception of space [MAR 99]. These relations are called qualitative. Distance and direction relations can also be expressed on a qualitative level. For the distance relation, the three values "exactly there", "close", or "far" are often used; for the direction relation, depending on the referential, various direction bases are available: cardinal (North, South, East, and West), user centric (in front, behind, left, right, above, and below). We will detail these two types of spatial relations currently available in GEOLIS: a quantitative relation (distance) and a qualitative relation (topology). These two relations are actually derived from entity geometry just like area and shape.

6.3.2.2.1. Distance

We can define $\mathcal{L}^{dist}$ as the logic of distance between entities. It creates a value domain matching the distance attribute. We define the distance between two geometries g_1 and g_2 as the minimal distance "as the crow flies" between a point p_1 in g_1 and a point p_2 in g_2. This is the definition of distance commonly used in GIS. However, other definitions are possible, and, for instance, we could have defined the distance between two geometries as the distance between their barycenter.

Since distance is a measure of length, the logic $\mathcal{L}^{dist}$ uses the same formulas and subsumption as the length logic $\mathcal{L}^{Length}$, but it matches the attribute distance. The logic $\mathcal{L}^{dist}$ allows us to define queries such as distance:(<=d) -> q_1, which select all the objects within a distance inferior to or equal to d of at least one object in q_1. This type of query is an elementary function in GIS which corresponds to the expression of a buffer zone [LAU 92], a region determined both by a set of origin objects (the entities in layer q_1) and by a radius d around these objects.

6.3.2.2.2. Topology

The spatial topological relations which we will simply call topological relations are binary relations qualifying the spatial position of an object relative to another. Examples of topological relations are the inclusion of an object in another, the connection of two objects, or even the overlapping of two objects. These relations are of a qualitative nature; they qualify a spatial organization by unvarying properties topological transformations such as rotation, translation, or scale changes. Modeling topological relations is a research field which has been greatly explored in the last 20 years [COH 97, EGE 89, RAN 92].

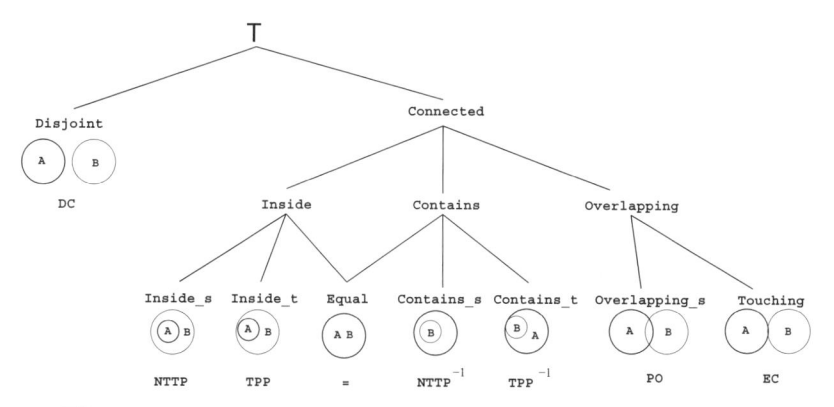

Figure 6.6. *A topological spatial relation taxonomy between two regions according to [WES 00]. The taxonomy leaves represent the eight basic relations of the RCC-8 model*

We have chosen to adopt a topological relation taxonomy suggested by Wessel, Haarslev, and Möller [WES 00] (see Figure 6.6). From the point of view of information searching, this taxonomy has the advantage of being relatively simple (12 relations), includes the eight basic relations of the RCC-8 topological model, and has easily understandable intermediate relations (connection, inclusion, and overlapping). This taxonomy is originally meant to represent relations between region pairs, but each of the eight basic relations is extended to any geometry couple (e.g. a point and a line).

The $\mathcal{L}^{Topo}$ precisely reflects this relation taxonomy. Each term of the taxonomy is a formula representing a specific topological relation, and a relation r_1 is subsumed by a relation r_2 if r_1 is a descendent of r_2 in the taxonomy. For example, the relation contains_t is subsumed by the relation connected. During the description of relations between object pairs, only seven of the eight basic relations are used. The relation disjoint is not expressed, and objects disconnected from a layer q_1 can be reached through the navigation link NOT connected -> q_1. The more generic relations can be used in queries or suggested by the system as navigation links. Finally, due to the symmetry between certain topological relations, we can see equivalences between different queries. For instance, the queries contains <- house ("something contained in a house") and inside -> house ("something inside a house") are equivalent due to the symmetry between the relations contains and inside.

6.3.3. *Use cases*

A GEOLIS user can know exactly what kind of information he/she is seeking and how to express this search as a query, or it can be the opposite and not have any clearly defined goal, and he/she might just let himself/herself be guided by the system in his/her exploration of the dataset. These two cases are the extremes of a *continuum* in information searches. We will show four use cases in this section, from the more directed to the less directed.

As we have done previously, we will base ourselves on the Milliau dataset to illustrate these use cases. Figure 6.7 shows the state of the interface when it opens the dataset[4]. At the level of the cartographic view, we can see that two noticeable zones of the island have been identified by two geometrical navigation links. The first is at the heart of the island and encompasses a set of buildings, and the second, on the north-west of the island, encompasses a ruin (according to the indications provided by the legend).

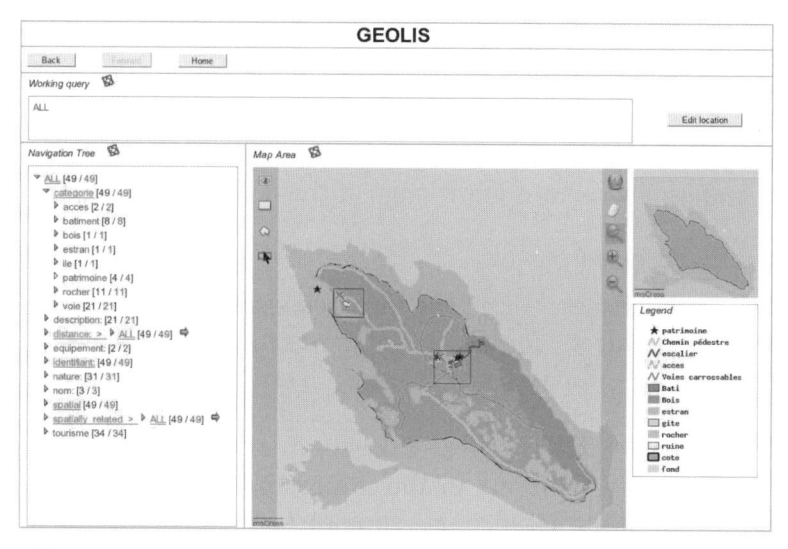

Figure 6.7. GEOLIS *interface before exploring the Milliau dataset. For a color version of this figure, see www.iste.co.uk/Bucher/innovgis.zip*

6.3.3.1. *Direct search*

A direct search is a search which directly formulates what we are searching for with a query, by typing it into the query view. For example, a user wishing to locate and

---

4 In the following screen captures, to improve visibility, we will not show the navigation bar in the background nor the right-hand side of the cartographic view (legend and situation map).

book lodging on the Milliau island can type the `gite` query. Let us immediately note that even this very simple query requires knowledge of the dataset, or luck. Indeed, the `gite` property could have been coded differently (for instance, `lodging:"gite"`) and there could be no gîte, only a camp site. After validating the query, the interface is updated: the map highlights three objects corresponding to the three gîtes on the island. Then since the user is part of a family of three, he/she refines his/her query into `gite AND capacity:3`. This time, there is no object meeting his/her query and he/she gets no answer. The following use cases aim to show the point of GEOLIS and more specifically of the navigation links, compared to a direct search which does not allow us to control the answer volume (too many or too few answers), and which presupposes a certain knowledge of the dataset.

6.3.3.2. *Targeted search*

As in the direct search, the targeted search aims to identify and locate objects corresponding to certain precise criteria. However, we suppose that the user has no knowledge of the dataset and should be guided in his/her search. In the initial view, the user sees a list of attributes matching different properties and relations in the index: for example, category, description, and distance. He/she starts by unfolding certain attributes to show more specific properties and finds, under the property `category`, the property `building` which is close to what he/she is looking for. By unfolding the `building` property, the index indicates that among the eight buildings there are three gîtes, one ruin, and four service buildings. He/she now only needs to select `gite` to refine his/her query (ALL ⤳ `gite`) and thus selects the three gîtes. The cartographic view shows that these three gîtes are one next to the other, meaning that location is not a discriminating criterion. By unfolding other properties, the user then finds that these three gîtes can essentially be differentiated by their name and capacity. By unfolding the property `capacity`, he/she finds that the available capacities are of four, five, or six people. He/she selects the property capacity:4 which is the closest match to what he/she is looking for. The index then tells him/her it is the gîte called "Molène".

We can see that the index plays two roles: it guides the user with relevant navigation links and provides him/her feedback on the objects selected. The first role guarantees correct queries which have answers. It also guarantees that each refining process is effective, i.e. discriminating. The second role increases the information fed back to the user and helps him/her choose in context. Thus, the user learned that there was no other type of lodging, that all the gîtes were gathered in one place, and that they all had a capacity of four or higher. All this feedback helps make the user confident about his/her choices and the results provided by the system.

6.3.3.3. *Exploratory search*

The exploratory search is based on the same mechanisms as the targeted search, but without any predefined goal. In our example, after having identified a gîte as a lodging, the user wonders what he/she might do or visit apart from this gîte. He/she starts by unfolding the relational formula distance:-> ALL, which brings up various distance intervals: for example, less than 50 m, between 100 and 200 m. To access objects which are less than 50 m from the gîte, he/she carries out a crossing on the relational increment distance:([0.0,49.99]) -> ALL, which brings him/her to the following query:

```
distance:([0.0,49.99]) <- (gite AND
capacity:4)
```

The views are updated and the index shows that among these objects are three service buildings, three points of interest (cultural heritage), a forest, and 12 roads. These objects are highlighted on the map. The user then decides to take a stroll

178   Innovative Software Development in GIS

from the gîte and wonders where these roads go. He/she starts by selecting the roads by refining the property road (see Figure 6.8) which brings him/her to the following query:

```
road AND distance:([0.0,49.99]) <- (gite
AND capacity:4)
```

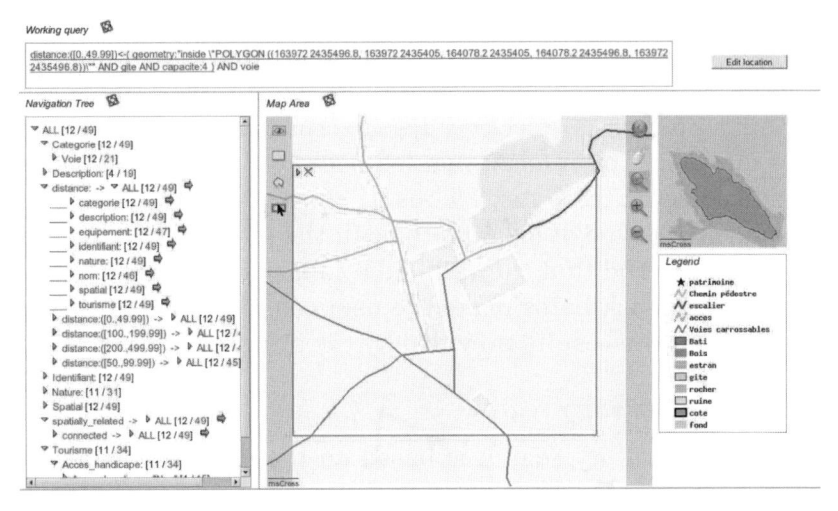

Figure 6.8. *Roads in the vicinity of the gîte (less than 50 m away). For a color version of this figure, see www.iste.co.uk/Bucher/innovgis.zip*

To know all the objects accessible from these roads, he/she can use topological relations between objects. By unfolding the relational increment spatially_related -> ALL, which corresponds to the most generic topological relation, he/she reaches the overlapping -> ALL increment, signifying "to have an intersection with something". By operating a crossing of this increment, he/she reaches objects intersecting roads that are also less than 50 m from the gîte, in the following query:

```
overlapping <- (road AND
distance:([0.0,49.99])
<- (gite AND capacity:4))
```

The index shows that these objects are other roads, one of the island's access points, and a forest. The user decides that the forest is an interesting destination and selects it by refining, leading to the following query:

```
forest AND overlapping <- (road AND
distance:([0.0,49.99]) <- (gite AND
capacity:4))
```

To identify which roads effectively lead to the forest from the gîte, the user considers the relational branch overlapping <-... and reverses the query. This causes the query to be re-centered on the roads, without losing any of the constraints accumulated throughout navigation (see Figure 6.9):

```
road AND overlapping -> forest AND
distance:([0.0,49.99])
<- (gite AND capacity:4))
```

The map now only displays two roads among the 12 that are close to the gîte. To fix his/her itinerary completely, the user looks for road discriminating properties in the index and finds that only one of the roads has disabled access. Since he/she has a child in a stroller, he/she then refines the query with the property disabled_access:"Yes" and the final query is:

```
road AND disabled_access:"Yes"
AND overlapping -> forest AND
distance:([0.0,49.99]) <- (gite AND
capacity:4))
```

The index then shows that this is a dirt road of about 500 m long, so the stroll will be a round trip of about a kilometer.

Obviously, this is only one possible scenario, since the goal of taking a walk was not present when the exploration started, but spontaneously appeared and progressively took shape according to the feedback provided by GEOLIS. Other

180   Innovative Software Development in GIS

scenarios could have unfolded centering around points of interest or service buildings.

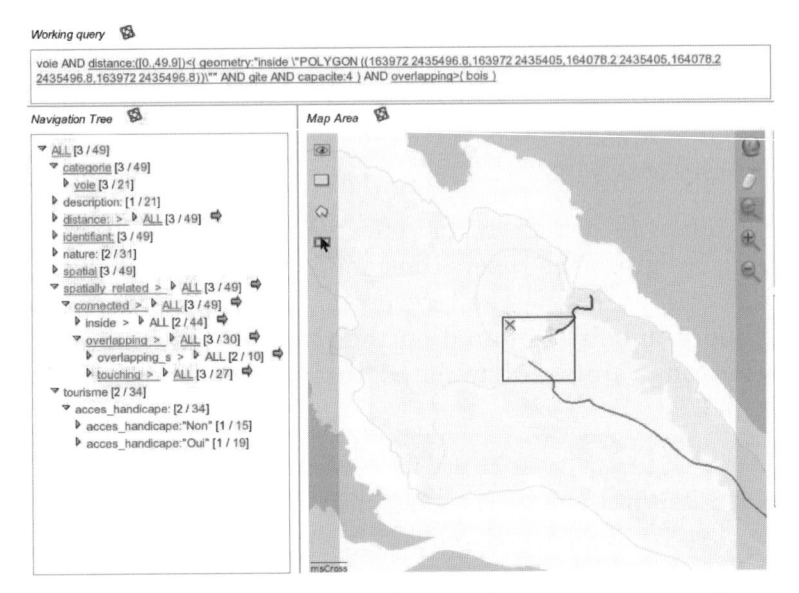

Figure 6.9. *Return on the "road" point of view after selecting the "forest" destination. For a color version of this figure, see www.iste.co.uk/Bucher/innovgis.zip*

6.3.3.4. *Knowledge search*

The knowledge search is again based on the same navigation mechanisms as previously mentioned. However, while targeted and exploratory researches had objects or groups of specific objects for research results (such as the gîte, the roads leading to the forest), knowledge searches aim for global knowledge of the island. The principle of discovering such knowledge is to select a subset of objects and read it in the index made up of the properties of these objects and relations toward other objects. The numbers matching the increments in the index allow us to narrow the associations between the query Q and the query X with quantifiers, for instance "there are n X among the Q", "all the Q are X", and "most of the X are Q". Such an exploration of the dataset on the Milliau island allows us to learn the following associations:

– The island has two main access points, roads, eight buildings, four points of interest (cultural heritage), a forest, and rocks.

– Among the eight buildings, there are three gîtes, four service buildings, and a ruin. Almost all of them have an area under 100 m^2, and the largest is the ruin at 282 m^2.

– Next to the buildings (less than 50 m away) are six of the eight buildings, three of the four points of interest but only three of the 11 rocks. We can thus deduce that buildings and rocks are in relatively separated zones and that two buildings are relatively far from the other buildings.

– According to the map, three points of interest are in the heart of the island and the fourth is North. The index provides us with the specific nature (such as a fountain) and indicates that they all have disabled access.

– The rocks have very unequal areas. Among these rocks, five are on the edge of the island (topological relation `inside_t`) while others are inland (topological relation `inside_s`).

– The roads provide access (topological relation `overlapping`) to the two access points for the island, to the forest, and of course are connected to other roads. Most are dirt, and some are gravel. The longest is 700 m long, but most are under 100 m long.

– Starting from an access point and repeatedly crossing the topological relation `overlapping`, we reach the other access point of the island, thus proving the existence of a path between them.

This list shows that without any prior knowledge and any precise goal, we can already learn a lot with a dataset, in this case about the Milliau island. We should also add that the above list does not reflect the thematic part of knowledge since the GEOLIS interface provides, in addition, the geographic distribution of every subset of objects considered (such as buildings and rocks).

6.4. Architecture

GEOLIS comes as a Web application with a client/server architecture. The client side serves as an interface for the user to query, navigate, and visualize virtual layers and matching information (increments and ratios). The server side covers logical context management. Starting with the geographic data in their original form, it provides different virtual layers when the client part requests them. As Figure 6.10 shows it, GEOLIS combines various technologies stemming from the field of LIS and Web mapping.

The heart of GEOLIS is made up of the logical file system LISFS [PAD 05b]. Geographic data are stored as files in a standard geographic data format (GML format) from which a set of objects is extracted. These objects have two components: their content in which is embedded all the information contained in their original layer and their logical description, which is all the logical descriptions which serve as criteria to organize and search the information. These thematic and spatial descriptions are automatically extracted from the original layer thanks to programs called transducers, and can be manually completed by the user.

A transducer is a program which must offer the following interface: (1) read on the standard input the file describing the layer to be loaded into the logical context of GEOLIS, and (2) write on the standard output a description line for every line read on the input file. For example, the following line matches the description of a gîte object on the Milliau island (the "/" serves as description separator):

```
/gite/capacity:3/name:"triagoz"/
disabled_access:"Yes"/...
...tourism:"dressed stone"/geometry:POLYGON
((...))
```

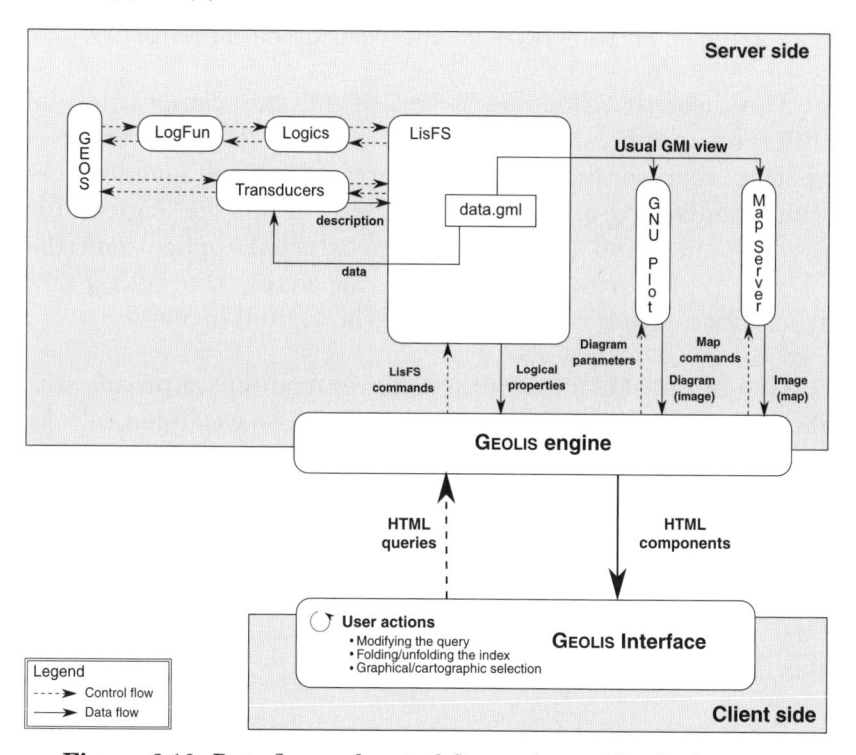

Figure 6.10. *Data flow and control flow exchanged by the* GEOLIS *interface and its different components server side*

GEOLIS is equipped with a GML transducer that allows it to accept encoded geographic data in that standard input. The GML transducer carries out this transformation of data into a description by relying on an XSLT spreadsheet. It can easily be adapted to other formats such as KML. More detail about the way transducers operate, including how they extract relational descriptions, is available in [BED 09].

The definition of thematic and spatial description language, as well as the reasoning modalities, requires specialized logical deduction *plugins*. These logical *plugins* can easily be built as a combination of elementary deduction engines supplied by the LOGFUN component library [FER 06]. It also can be based on the GEOS geometrical processing library.

The user interface is a result of the composition of different graphical components, each of them offering a specific representation of the current virtual layer. These components are either directly built from the logical file system (index and query) or require external applications (the MAPSERVER cartographic engine for maps, the GNUPLOT tracer for a graphical projection of the virtual layer).

One of GEOLIS's advantages is its modular approach: the data transducers and logical *plugins* can be extended to take new data formats, new value fields into account. In the same way, the interface organization can be personalized. This allows an application designer to adapt the tool to the data manipulated and the exploration and visualization needs.

6.5. Users and developers

GEOLIS does not have a user community yet since it is very recent and it is still considered a prototype. It has however already been applied to a set of real data which interests the RESO geographers. This is a dataset compiled by the Research Institute for Development (IRD) over decades, concerning the spread of different rodent species in Sudanese-Sahelian Africa. It has over 20,000 individuals with the time and place of the sample taken and their characteristics: all in all 92 attributes, such as gender, species, and weight.

Using GEOLIS to explore this dataset immediately revealed anomalies which were almost invisible in the original table due to its size. Sometimes it is values which are incompatible with the attribute to which they are matched, such as `Sex:49`; sometimes one value has different shapes, such as `Sex:"M"` and `Sex:"m"`. These anomalies can be explained by the fact that all the data were manually input into the table. For example, reading the index under the `country` attribute reveals that half of the samples were taken in Senegal, while this country only covers a small portion of the Sudanese-Sahelian area. Some navigation stages allowed us to learn that 85% of the rodent samples taken in the savannah were taken in 2000. The IRD biologists acknowledged that GEOLIS was better than other tools in detecting such anomalies and biases. More generally, they believe that GEOLIS "can help them validate or refute hypotheses explaining the presence of rodents by observing correlations between the attributes of a selection of rodents".

From a developer's point of view, GEOLIS's extensibility mainly lies in the extensibility of LISFS, a logical file system. Indeed, GEOLIS is itself a LISFS extension that keeps the same extension abilities as LISFS. The extension points shared with LISFS are, on the one hand, the logics which come into play in the data description and query language and, on the other hand, the transducers which convert data from their original format (such as GML) into the format used by LISFS. These extension points were used by GEOLIS to define the logics and transducers corresponding to geographic data. Another extension point, unique to GEOLIS, is the Web interface in which new components can easily be integrated. We have, for example, added graphics generated by GNUPLOT as additional dynamic views on the data.

6.6. Conclusion

To conclude, we will provide a brief overview of our perspectives for GEOLIS. They also reveal its current limits in processing geographic data. They focus on query language expressiveness and navigation, data visualization, and immediate data edition in GEOLIS. When it comes to expressiveness, we would like to include a maximum of geometric operations classically available in GIS, and more specifically all forms of data aggregation. To visualize the results of these aggregations, we can take inspiration from what is happening with OLAP cubes [COD 93] and their spatial extension, SOLAP. From a more pragmatic point of view, we would like to merge our different implantations by separating a LIS core playing the role of server (storage, access, and data editing), a Web client interface (visualization and navigation), and a set of reusable and extensible components to adapt it all to different applications.

6.7. Bibliography

[BED 09] BEDEL O., GEOLIS: un système d'information logique pour l'organisation et la recherche de données géolocalisées, PhD Thesis, University of Rennes 1, January 2009.

[COD 93] CODD E.F., CODD S.B., SALLEY C.T., *Providing OLAP (On-line Analytical Processing) to User-Analysts: An IT Mandate*, Codd & Date, San Jose, USA, 1993.

[COH 97] COHN A.G., "Qualitative spatial representation and reasoning techniques", *KI '97: Proceedings of the 21st German Conference on Artificial Intelligence*, Springer-Verlag, LNAI 1303, Freiburg, Germany, pp. 1–30, 1997.

[EGE 89] EGENHOFER M.J., "A formal definition of binary topological relationships", *FODO '89: Proceedings of the 3rd International Conference on Foundations of Data Organization and Algorithms*, Springer-Verlag, London, UK, pp. 457–472, 1989.

[FER 04] FERRÉ S., RIDOUX O., "An introduction to logical information systems", *Information Processing & Management*, vol. 40, no. 3, pp. 383–419, 2004.

[FER 06] FERRÉ S., RIDOUX O., Logic functors: a toolbox of components for building customized and embeddable logics, Research Report no. RR-5871, IRISA, March 2006.

[FER 09] FERRÉ S., "Camelis: a logical information system to organize and browse a collection of documents", *International Journal of General Systems*, vol. 38, no. 4, pp. 379–403, 2009.

[HER 06] HERRING J.R., OpenGIS implementation specification for geographic information (06-103r3), Open Geospatial Consortium, 2006.

[LAU 92] LAURINI R., THOMPSON D., *Fundamentals of Spatial Information Systems*, Academic Press Limited, London, UK, 1992.

[MAR 99] MARK D.M., "Spatial representation: a cognitive view", MAGUIRE D.J., GOODCHILD M.F., RHIND D.W., LONGLEY P., (eds), *Geographical Information Systems: Principles and Applications*, 2nd ed., John Wiley & Sons, Hoboken, pp. 81–89, 1999.

[OBJ 01] OBJECT MANAGEMENT GROUP, OMG Unified Modeling Language Specification Version 1.4, September 2001.

[PAD 05a] PADIOLEAU Y., Logic File System, un système de fichier basé sur la logique, PhD Thesis, University of Rennes 1, February 2005.

[PAD 05b] PADIOLEAU Y., RIDOUX O., "A parts-of-file file system", *USENIX Annual Technical Conference*, General Track (Short Paper), 2005.

[RAN 92] RANDELL D.A., CUI Z., COHN A., "A spatial logic based on regions and connection", *KR'92. Principles of Knowledge Representation and Reasoning: Proceedings of the 3rd International Conference*, Cambridge, USA, Morgan Kaufmann, San Mateo, CA, pp. 165–176, 1992.

[WES 00] WESSEL M., HAARSLEV V., MÖLLER R., "Visual spatial query languages: a semantics using description logic", *Diagrammatic Representation and Reasoning*, Springer, Berlin, 2000.

Chapter 7

GENEXP-LANDSITES: a 2D Agricultural Landscape Generating Piece of Software

7.1. Introduction

The GENEXP-LANDSITES software is a two-dimensional (2D) random agricultural landscape generator, first developed in 2003 as part of a study on the coexistence of genetically modified crops with traditional crops. It uses tools stemming from spatial statistics and algorithmic geometry. Its main goal was to provide agricultural landscape maps to test the role of certain characteristics of these landscapes faced with transgene dissemination. Indeed, until the beginning of the noughties, agrobiologists worked with dissemination models that were developed at the level of a few fields and lacked any real data allowing them to scale up to the level of a landscape, even though various studies had shown it

Chapter written by Florence LE BER and Jean-François MARI.

was necessary [ANG 02]. On a practical level, GENEXP-LANDSITES offers the agrobiologist user a set of methods to generate and describe landscapes – field pattern and land use – which are then used as input for gene flow models.

The following chapter follows the main outline of our book: a context description (section 7.2), a presentation of the major functionalities of GENEXP-LANDSITES, and the description of a case study (sections 7.3 and 7.4), the architecture details (section 7.5) and a few elements about the user and developer community (section 7.6).

7.2. Context

Landscape[1] simulation has been an important field of research for over 20 years in landscape ecology, and, more recently, in agronomy. The goal of landscape ecology is to study the role of landscape structures faced with ecological dynamics [TUR 91]. Simulation methods can be divided into three groups: geostatistical models, neutral landscape models, and explicit process models [SAU 00]. Geostatistical models are based on spatial data interpolation methods [GOT 96]. In explicit process models, the landscape is the result of modeled ecological processes (dispersion, competition, etc.). On the contrary, neutral landscape models (NLM, as described by [GAR 87]) provide random landscape structures which can serve as references to compare with real landscape, or which allow us to test the effect of certain landscape structures, such as, for instance, the habitat fragmentation, on ecological processes [GAR 87, GAR 91, WIT 97]. The NLM are essentially based on *raster* approaches, land use (often limited to two categories) is randomly allocated to pixels which are then grouped to create

---

[1] Landscape means here a 2D mosaic representing an area going from a few square kilometers to several dozen square kilometers.

fragments of space with similar use. To obtain more "realistic" landscapes with various categories, classification [SAU 00] or fractal [HAR 02] approaches have also been developed. Approaches based on tessellations were recently suggested to simulate geometrically shaped landscapes, forest landscapes, or agricultural landscapes [GAU 06a, GAU 08].

Landscape models were frequently used to help manage forests [KUR 00, LIU 98]. In the fields of agricultural land management and agronomy, these models are less common. Spatial models have notably been used to simulate crop distribution within a farm or a piece of land [CAR 02, LEB 98] to understand the impact of agriculture on ecological issues (such as drinking water pollution) or to plan more favorable spatial organizations. Closer to what has been done in ecology, some studies have explored the link between agricultural practices (crop rotation, land consolidation, etc.) and ecological processes thanks to explicit process models [AVI 07, GAU 06b]. All these models are based on real data such as spatial distributions of fields, physical characteristics of landscapes (soil conditions, slope, etc.), or crop distribution on the studied zone. Beside a few recent works [GAU 08], there is no neutral model in those fields such as the ones developed in landscape ecology. These models are interesting for various reasons. On the one hand, real data are not always available or can be too specific, thus restricting the scope of the model's results. On the other hand, prospective studies can be based on new landscape configurations which did not exist before. Finally, neutral landscape models can be used to test a process model's sensitivity to agricultural land spatial variability.

Beyond the structure simulation, agricultural landscape models tackle the simulation of land use in various ways. In most landscape models, land use is randomly allocated on the modelized area according to a law of probability: for example, in binary landscapes, each plot has a probability of being planted with a specific crop that depends both on its area and a value p matching the crop. When various land uses are modeled, each land use is matched to a probability p_c so that $\sum_c p_c = 1$. There have been various methods suggested to improve the purely random approach. In [GAU 06a], the authors introduced a Gibbs process to model interactions between pairs of adjacent fields. The landscape mosaic can be more or less heterogeneous depending on the parameter values of this model. Other processes can be used, based on the statistical study and spatiotemporal pattern characterization of real agricultural mosaics [CAS 07, CAS 08]. Moreover, stochastic rules of adjacency can be learnt from real sets of data, as [MAR 06] suggested it. In this study, the hidden Markov chains were implemented on land use spatiotemporal data. Deterministic approaches can also be used, especially to study the effects of decision rules or evolution scenarios of land use [CAR 02, GAU 06b].

In France, we can currently see a rise in landscape models to deal with various issues. For example, researchers participating in the RECORD[2] platform are considering developing a 2D landscape scale to represent the landscape distributions of crop systems. Other models focus more on modeling the physical process and include soil and slope characteristics – often based on real data, which can be distorted: a virtual landscape model is described in [SOR 08], for instance. We can also mention the APILand tool developed by the INRA SAD-Paysage unit in Rennes [BOU 10]. The GENEXP-LANDSITES model presented here has been developed to study the coexistence of genetically modified organism (GMO) crops with traditional crops [ADA 07, LEB 09]. Agrobiologists wanted to study in more detail the

---

2 http://record.toulouse.inra.fr/

role of agricultural landscape structure in the dissemination of transgenes (originating from GMO) thanks to wind, regrowth or mechanical transportation. Moreover, they tried to characterize landscapes in which the risk is limited [COL 09b, LAV 08]. The elaboration of GenExP-LandSiTes thus happened originally within the framework of the research project "Modeling transgene dissemination at the scale of agricultural landscapes" (answering the call to tender for "OGM Impact" put out by the French Research Ministry between 2003 and 2006) and required a panel of experts (in biology, statistic, agronomy, and computer science). Computer science students implemented the first versions [DEL 04, GUE 03]. GenExP-LandSiTes's Gnu Public License was then filed with the Program Protection Association[3]. Further releases were carried out thanks to student internships with no specific funding [KAL 07, MOR 06]. However, the constant interest expressed by agrobiologists recently allowed us to obtain funding within an INRA–INRIA call to tender. The PayOTe project "Agricultural land and landscape models to study the agro-ecological process" (2009–2010)[4] brought together a set of INRA and INRIA researchers to think about these issues. These reflections led to new development in 2009 and 2010, to strengthen and expand GenExP-LandSiTes's functionalities [BRO 09, KOS 10].

7.3. Major functionalities

The GenExP-LandSiTes software is not a geographical information system but a 2D virtual agricultural landscape generator. This means it does not manage spatial information about the landscapes – even if it can use certain data

---

[3] Effective filing in 2006, holders: INRIA, University of Paris Sud, ENGEES, University of Nancy 2, INRA, INA Paris-Grignon.
[4] Project renewed in 2011 as "Mining and simulating 2D landscapes – Configuration and composition".

extracted from real landscapes – but it allows us to create more or less realistic field pattern maps and stands as a complementary tool for GIS. It has two main functionalities: field pattern simulation and cropping pattern simulation. It relies on coupling with the R piece of software to offer more functionalities, concerning point process simulation and spatial analysis. We will thus successively present: (1) point generation, which requires R's libraries, (2) field pattern simulation, based on space tessellation methods, (3) cropping pattern simulation, and (4) post-production and spatial analysis.

7.3.1. *Point generation*

To generate points (or tessellation seeds), GENEXP-LANDSITES relies on the libraries of the piece of statistical software R and more specifically on its spatial statistics library, SPATSTAT [BAD 05]. There are more than one option possible: random approaches based on neutral processes (Poisson processes) or on processes fitted to the characteristics of a real landscape. Thus the barycenters of the fields of a real landscape can be used as seeds for virtual landscapes. They can also be used to asses a distribution model of plot barycenters [ADA 07], allowing us to control a certain variability in their number and location in virtual landscapes (Figure 7.1).

7.3.2. *Field pattern simulation*

The main goal of GENEXP-LANDSITES is to offer different field pattern generation algorithms. We are showcasing two here: the Voronoï diagrams and a random rectangular tessellation.

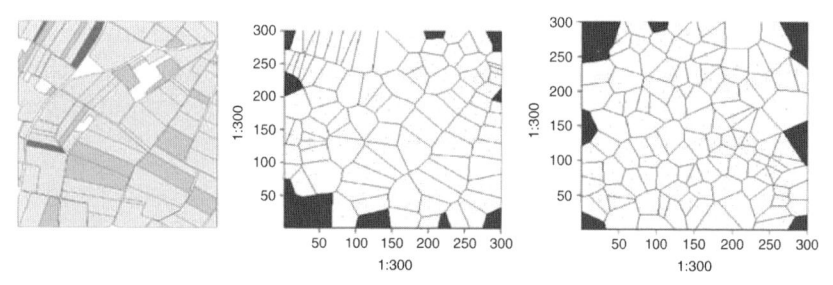

Figure 7.1. *Three field patterns: (left) real landscape, (middle) simulated – Voronoï diagrams – landscape based on real barycenters, (right) simulated landscape based on simulated barycenters*

7.3.2.1. *Voronoï diagrams*

A Voronoï diagram is a partition of the Euclidean plane $\mathbb{R}^2$, generated from a set of points E, called "sites" or "seeds". Each seed g_1 matches an element p_1 of this partition, defined as the subset of the points in the plane which are closer to g_1 than to any other seeds in E [OKA 00]. The partition is thus built around as many polygons as there are seeds. These polygons are called "Voronoï polygons" and are convex. Their edges are thus made of the points which are equidistant from two seeds (Figure 7.2(a)).

a) Voronoï diagrams   b) Rectangular tessellation

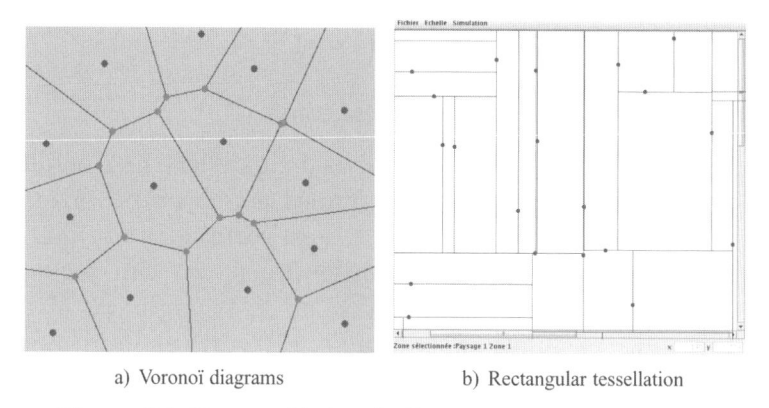

Figure 7.2. *Two tessellations built on point seeds: a) Voronoï diagrams and b) rectangular tessellation with T vertices*

On a practical level, GENEXP-LANDSITES first builds the Delaunay triangulation by using the 3D convex hull algorithm (see Algorithm 7.1, and also [OKA 00]) and then determines the matching Voronoï diagram. The algorithm insertion is derived from [ORO 98].

Algorithm 7.1 3D Convex hull

Input: a set $P = \{p_1, p_2, \ldots, p_n\}$ of n seeds whose coordinates are (x_i, y_i) for any $i \in [1, n]$
Output: Delaunay triangulation on the set P
Process:
Step 1: build the set $P^* = \{p_1^*, p_2^*, \ldots, p_n^*\}$ in dimension 3 in which the two first coordinates of p_i^* are those of p_i and the third coordinate is $x_i^2 + y_i^2$.
Step 2: build the convex hull $C(P^*)$ of P^* in a 3 dimensional space.
Step 3: project all the lower bifaces of $C(P^*)$ in the original 2 dimensional space on a parallel with the third axis; flip the resulting diagram.

Each polygon is then identified as a field of the simulated agricultural landscape. The result can be easily linked to the structure of a real landscape; we can, for example, obtain the seeds from the barycenters of the real landscape fields.

7.3.2.2. Random rectangular tesselation

A rectangular tessellation allows us to divide the Euclidean space into jointed rectangular shapes which do not overlap. By eliminating the simple case where rectangles are defined by two orthogonal sets of parallel lines, we can focus on a tessellation where the vertex of a triangle is always on the side of another rectangle (T vertices and not X vertices, see Figure 7.2(b)). GENEXP-LANDSITES implements a method described in [MAC 96]. The algorithm principle (see Algorithm 7.2) is to generate the sides of the rectangles from a set of points: from each point start two opposed segments which are parallel to the x-axis or the y-axis. When two segments meet, the longer one stops. We then have a number of rectangles equal to the number of points minus one.

Algorithm 7.2 Rectangular tessellation

Input: a set $P = \{p_1, p_2, \ldots, p_n\}$ of n seeds; each p_i has a direction d_i, vertical or horizontal, with probabilities of $1/2, 1/2$.
Output: a list of n couples of points (radius extremities).
Process:
Step 1: for each couple (p_i, d_i), the adjacent d_j orthogonal to d_i are identified; they are potential blockers.
Step 2: for each couple (p_i, d_i), a radius is drawn along the direction d_i until it reaches its closest blocker.
Step 3: the isolated (unfinished) radii are dealt with sequentially. The current RC radius is prolonged towards its closest blocker, RB. There are then three possibilities:
(i) RB is already finished (on another line); RC is prolonged until the next blocker;
(ii) RB has not reached RC's line: RC is then suspended and RB becomes RC (the current radius);
(iii) RB went beyond RC's line: RC ends on RB and leaves from the list of isolated radii.

Variations on this algorithm have also been implemented by playing on the axis direction to obtain diversely oriented rectangles or non-rectangular parallelograms [KOS 10]. When it comes to the link between this type of tessellation and landscapes, we can see that the initial points are placed on any side of the rectangles. There is thus no obvious link between the tessellation seeds and remarkable points (barycenters, field vertices) of a landscape.

7.3.3. *Cropping pattern simulation*

GENEXP-LANDSITES allows us to randomly allocate land use to simulated fields, on the basis of a distribution of probabilities representing a cropping pattern (such as 50% corn, 30% wheat, etc.). GENEXP-LANDSITES also allows us to simulate the evolution of this cropping pattern over time, and the succession of crops on the different fields. The crop sequences are represented with Markov models (MM) or hidden Markov models (HMM) in two different ways. The models can be manually built as in [CAS 08], or we can obtain them thanks to sets of real data through the CARROTAGE software [LEB 06]. This stochastic data mining software works on temporal sequences of agricultural land use and represents these sequences with an HMM made of two states: a container state (written as "all" in Figure 7.3) representing a land use distribution similar to a cropping pattern, and a "Dirac" state in which there is only one type of land use (see "wheat", "corn", "rapeseed" in Figure 7.3).

Figure 7.3. *The different states of an HMM representing sequences dominated by rapeseed, wheat, corn*

7.3.3.1. *Stationary method*

This method uses transitional mean probabilities (also called *a priori* probabilities) between the states of an HMM which has already been learnt from real data gathered over an agricultural land [MAR 06, LEB 06]. In a first time t, GENEXP-LANDSITES allocates crops to the fields according to a given distribution which represents crop rotation. In a second time $t + 1$, land use is simulated by using the *a priori* transitional probabilities between crops calculated for this HMM with CARROTAGE. Land use is obtained through a random selection based on three multinomial distributions matching the transitions and distributions. The process can be repeated over n periods of time to simulate stable successions over time.

7.3.3.2. Taking into account succession changes

This method uses the diagram representing probabilities over time (also called *a posteriori* probabilities) of the transitions between states over a given period of time. Figure 7.4 shows a diagram for an open plain of cereal crops which is dominated by a 4-year crop rotation cycle of rapeseed–wheat–sunflower–wheat. This figure shows an increase in wheat monoculture since 1996 for which the HMM *a priori* probabilities do not account. The HMM used (see Figure 7.3) has the following individualized states: sunflower, winter grain (wheat), rapeseed, pastures, corn, not surveyed, or constructs. The state noted as "?" corresponds to the container state covering all the other land uses. Thanks to these *a posteriori* probabilities, we can simulate successive land uses with the same dynamics and over the same period of time as a given region by using multinomial distributions similar to the method described earlier.

7.3.3.3. Future changes

We are currently working on two major changes: (i) spatial allocation of land use with a Gibbs process similar to the one suggested in [GAU 06b], (ii) temporal and spatial allocation – to keep consistency among the landscapes simulated over time – thanks to classified HMM [LAZ 09].

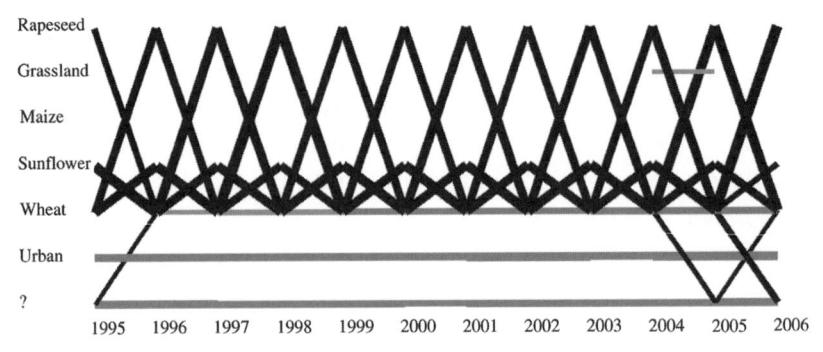

Figure 7.4. *Temporal evolution of crop successions (dominated by rapeseed, wheat, sunflower) with the a posteriori probabilities given by an HMM*

7.3.4. Post-production, spatial analysis, and formats

7.3.4.1. Post-production

GENEXP-LANDSITES allows for a certain number of post-productions on the generated landscapes. We can mention, among others:

– diminishing fields that spill over the borders of the landscape (especially for Voronoï diagrams);

– deleting segments that are too small (point merger, see Figure 7.5);

– deleting fields that are too small or merging them with an adjacent field.

7.3.4.2. Spatial analysis

As for the spatial analysis, we have basic functions: field number, field area, number of sides, perimeter, barycenter, and shape index. The matching statistics are displayed as a histogram for a landscape and/or a boxplot if we wish to compare landscapes.

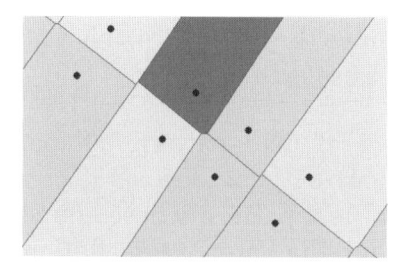

Figure 7.5. *Post-production: aligning segments by merging points (left, initial Voronoï diagram, right, after post-production)*

Using the R's SPATSTAT library [BAD 05] enables us to calculate other parameters, such as the distance of barycenters to their nearest neighbor (see below, in the case use).

7.3.4.3. *Formats, import, and export*

GENEXP-LANDSITES was originally devoted to producing maps for specific software. It can therefore manage these specific formats as well as GIS (*shapefile* format), image formats and XML formats.

7.4. Case uses

The GENEXP-LANDSITES software currently generates field pattern maps to be used by two gene flow models dealing with different crops. Colbach [COL 09a] is a multiannual model allowing us to predict the level of adventitious presence in rapeseed crops (e.g. the level of GMO in non-GMO crops or the level of erucic acid in varieties 00) to a field pattern made of tilled land and spaces outside the fields such as roadsides, a sequence of crops on each field and associated tillage techniques. The whole set is defined by the user at the model input. Angevin *et al.* [ANG 08] simulates gene dissemination of corn in a spatially heterogeneous space at the scale of a full crop year. It does not take the crop sequences and grain transfers into account, since they are not persistent from one crop year to the other in European climates.

The aim is to study the effect of field pattern (shape, location, and field area distribution) on gene flow, or it can operate on real field pattern or on field pattern whose shape will have been modified (e.g. by using the Voronoï tessellation on real field barycenters) or the shape and location of fields (for instance, by using a Voronoï tessellation on random seeds or a rectangular field pattern). Figure 7.6 shows a chain of tools and data required to implement such a study. Later on, GENEXP-LANDSITES can be used to study other spatialized ecological processes.

We will describe below in more detail the main stages of a study carried out with, and based on, real landscape data and on a set of simulations generated in GENEXP-LANDSITES. Detailed studies are described in [ADA 07, COL 09b, LAV 08, LEB 09].

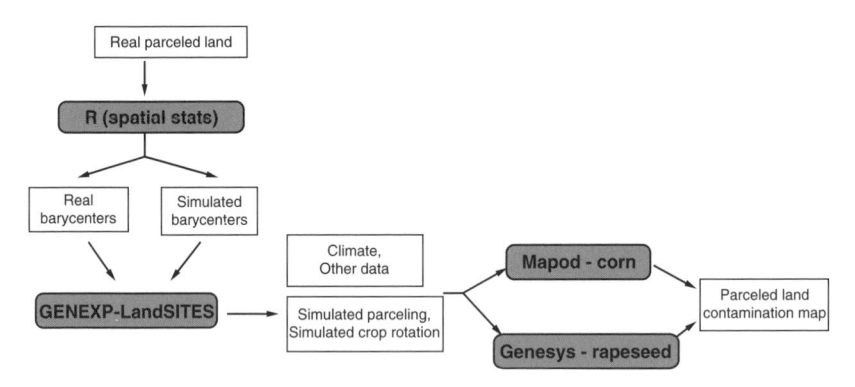

Figure 7.6. *Chain of different models and data to simulate the flow of genes in an agricultural landscape*

Suppose datasets from real field patterns are available. The biologist can use them as input data for the gene flow model. However, his/her goal is to study the effects of the variation of certain characteristics of these landscapes on gene flow. He/she thus uses GENEXP-LANDSITES to create simulated field patterns with characteristics close to the real field patterns.

Let us, for example, consider the original field pattern in Figure 7.1 (left): it is a field pattern of a corn zone of the Alsatian plain, its area is of 1,500 m by 1,500 m; there are 100 fields and their area is average and variable (2.05 ha ± 1.84), with a rather elongated shape (Shape Index[5] = 1.53 ± 0.28). Finally, the fields are rather narrow since the distance between a barycenter and its closest neighbor is on average 92.6 ± 39 m.

Using GENEXP-LANDSITES, the biologist can then play with different parameters to simulate landscape that are more or less similar to the original landscape.

– seed choice: (i) original seeds (real landscape barycenters), (ii) simulated seeds based on the real landscape barycenters, (iii) random seeds;

– tessellation choice: (i) Voronoï diagrams, (ii) rectangular tessellation;

– cropping pattern choice: (i) random allocation, (ii) use of a stochastic model learnt from the data of the real agricultural landscape.

By choosing to go for the (seeds: ii) (tessellation: i) (crop rotation: i) assembly, we can generate landscapes which have the following characteristics (means for nine repetitions): field number: 102.9 ± 11.2; average area: 2.29 ha ± 0.28 (average of intra-landscape standard deviations: 1.01); average shape: 1.03 ± 0.01 (average of the intra-landscape standard

---

[5] I_S = perimeter/$4\sqrt{\text{surface}}$ varies between 0.9 (for a circle), 1 (for a square), and 1.7 for a rectangle whose length is nine times its width.

deviations: 0.11); distance between the (closest) neighboring barycenters: 112.09 m ± 6.88 (average of the intra-landscape standard deviations: 29.11). We can note that if the number and average area[6] of the fields remain reasonably similar, the distance between barycenters and the shapes, however, are less variable and larger (for the distances), more compact (for the shapes) than on the original field pattern. These results are directly linked to the chosen tessellation, since Voronoï diagrams create convex compact shapes and maximize the distance between barycenters during the design phase. We can see two field patterns obtained with the chosen cropping pattern (three crops, occupying respectively 60%, 10%, and 30% of the land) in Figure 7.7.

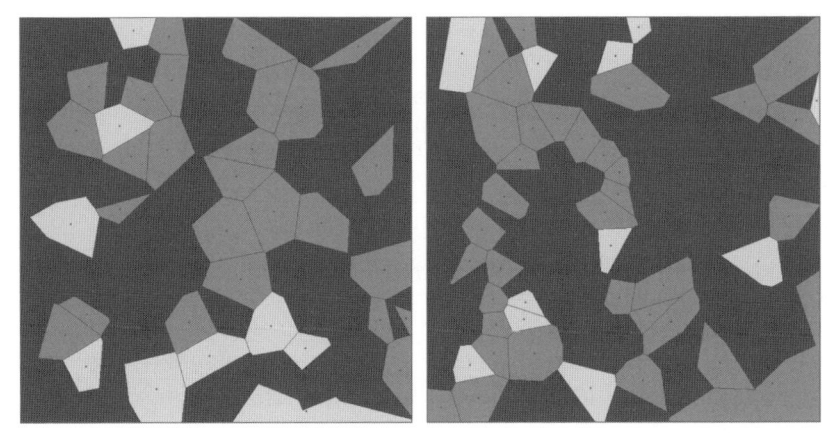

Figure 7.7. *Two simulated field patterns (Voronoï diagrams) with random crop rotations*

7.5. Architecture

The general architecture is shown in Figure 7.8.

---

6 In the simulation, the polygons fill the whole landscape, whereas the real landscape has holes (fallow grounds, forests, etc.): therefore, if there is an identical number of simulated fields as there are real landscape fields, the simulated fields will be larger.

7.5.1. *The application* `Core`

This class (see Figure 7.8) contains the list of landscapes thanks to the object `LandscapeManager`. The latter has all the information to allocate a new identifier to landscapes, and manages the list of landscapes. Thanks to this list, we can run different simulations over a few years. The `Core` is also the input for R management. It is indeed in this class that the `R-Manager` is placed, a general class which allows the instancing of classes enabling one to interact with the R application. Here also are usually inserted the inputs toward tessellation and I/O plugins.

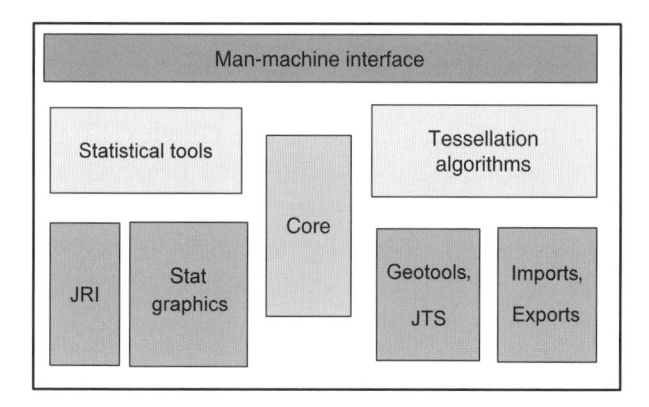

Figure 7.8. *Architecture*

7.5.2. *Separating graphical classes from business classes*

Graphical displays are handled by a `Panel`-type object. A `LandscapeCreator` type object creates a link between the graphics and the landscape. Some mouse functionalities, such as its state as well as its previous state, are externalized into a specific class.

7.5.3. *The plugin system*

GENEXP-LANDSITES uses a plugin system to represent all the tessellation and landscape I/O algorithms. Its *modus operandi* is simple. The plugin classes do not need to know their context of use. They are all stored in a class that serves as PluginManager and is integrated in the Core application. The application builds its graphical interface depending on the plugins found when it boots up. For example, the export menu is dynamically created and depends on the loaded export plugins.

7.5.4. *Interface*

The user is led to successively manipulate different menus according to how he/she wants to use GENEXP-LANDSITES: it can be an exploratory use (for instance, visualizing the different type of landscapes that can be obtained) or a systematic use (generating a set of landscapes from a set of parameters).

The first step is choosing the size of the landscape to be generated, then to outline the zones. Then, for each zone, the user has access to a *landscape* menu (Figure 7.9) which allows him/her to choose:

– a seed generation method;

– a land use distribution;

– a tessellation method.

Then to access the post-production and embellishment (e.g. choosing the land use colors) and the statistics about each zone in the generated landscape (see Figure 7.10).

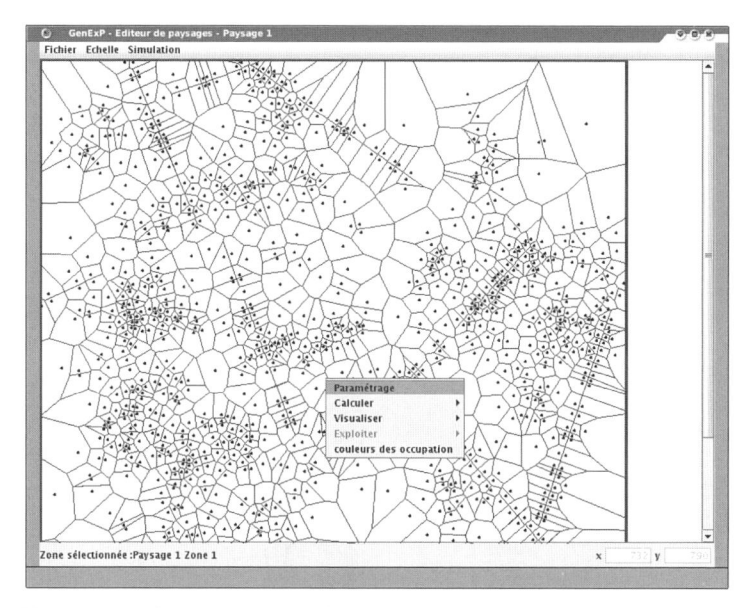

Figure 7.9. GENEXP-LANDSITES *interface: landscape menu (one zone, Voronoï tessellation with groups and aligments of seeds)*

7.6. Communities

GENEXP-LANDSITES was used for different studies carried out by INRA biologists who were interested in transgene flows [COL 09b, LAV 08] in collaboration with statisticians. Agronomists and other biologsts also expressed interest in this tool and are an active part of the reflection that is currently being carried out on landscape and agricultural land models (see section 7.2)

The contributors are mainly computer science students from degree and master level courses. The different versions of GENEXP-LANDSITES were thus developed during almost 10 successive internships or successive short contracts that took place between 2003 and 2010.

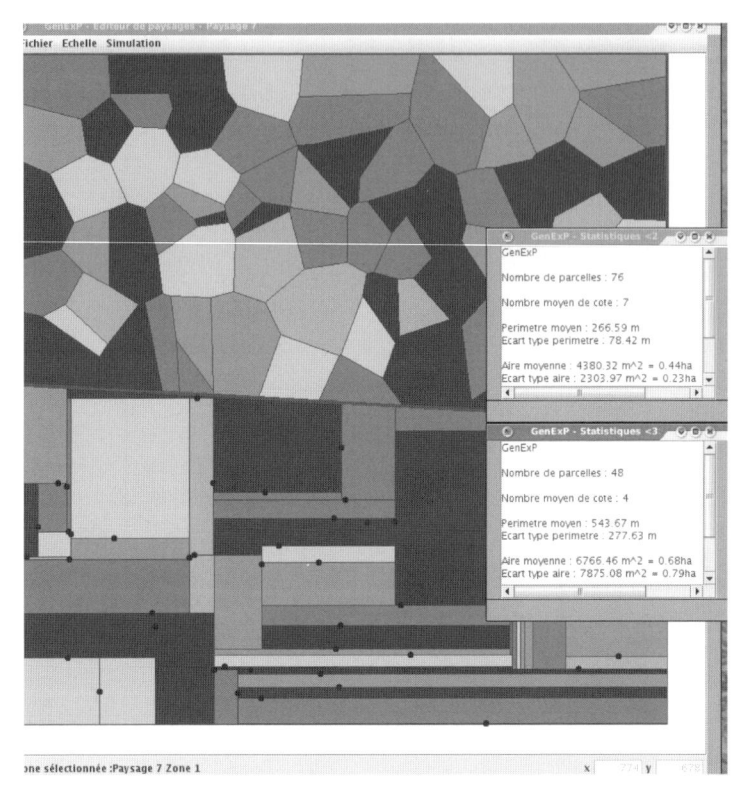

Figure 7.10. GenExP-LandSiTes *interface: landscape menu, statistics (field number, average perimeter and area) display on the different landscape zones (two zones, Voronoï and rectangular tessellation with random seeds)*

The main unit holders were at the beginning the LORIA-INRIA Lorraine, a computer science and applications research laboratory[7] in Lorraine, and the ESE (Environment, Systematics, and Evolution)[8] laboratory. Today, due to the change in the status of the personnel involved in GenExP-LandSiTes's development and due to the revamping of

---

7 CNRS, INRIA and Nancy Universities, http://www.loria.fr
8 U. Paris XI, CNRS, AgroParisTech, http://www.ese.u-psud.fr/

various organizations and laboratories, the main unit holders are the INRA, ENGEES[9], and the INRIA Nancy Grand Est[10].

The development of GENEXP-LANDSITES was funded by the French Research Department (answering the call to tender for "OGM Impact"), then by the teams' own funds, and more recently by the INRA within the PAYOTE project "Landscape and agricultural land models to study the agro-ecological process" (INRA-INRIA call to tender). These sources of funding did not give us the possibility of contracting an engineer, but a doctoral candidate – working within the INRA-INRIA framework agreement – started his/her dissertation on this topic in October 2010.

7.7. Conclusion

The development of GENEXP-LANDSITES should continue in a wider context. We will notably focus on implementing other tessellation techniques, on simulating other types of seeds (points, segments, and shapes). The spatial analysis possibilities should also be improved to offer users a complete toolbox going from a virtual landscape simulation to their analysis. Finally, we will try to use GENEXP-LANDSITES for new agro-environmental applications to test its use and strengthen the interest it presents.

7.8. Acknowledgments

We would like to thank all the people (researchers and interns involved in the projects "Modeling transgene dissemination at the scale of agricultural landscapes" and

---

9 http://engees.unistra.fr/
10 http://www.inria.fr/nancy/

"Landscape and agricultural land models to study the agro-ecological process") who took part in developing GENEXP-LANDSITES since 2003. The original field pattern data were provided for the project by the AUP (Single Payment Agency) and the IPSC, the Institute for the Protection and Safety of Citizens (Common Research Center).

7.9. Bibliography

[ADA 07] ADAMCZYK K., ANGEVIN F., COLBACH N., LAVIGNE C., LE BER F., MARI J.-F., "GenExP, un logiciel simulateur de paysages agricoles pour l'étude de la diffusion de transgènes", *Revue Internationale de Géomatique*, vol. 17, no. 3–4, pp. 469–487, 2007.

[ANG 02] ANGEVIN F., COLBACH N., MEYNARD J.-M., ROTURIER C., Analysis of necessary adjustements of farming practices, Scenarios for Co-existence of Genetically Modified, Conventional and Organic Crops in European Agriculture, Technical Report Series, EUR 20394 EN, Joint Research Center of the European Commission, 2002.

[ANG 08] ANGEVIN F., KLEIN E.K., CHOIMET C., GAUFFRETEAU A., LAVIGNE C., MESSÉAN A., MEYNARD J.-M., "Modelling impacts of cropping systems and climate on maize cross pollination in agricultural landscapes: the MAPOD model", *European Journal of Agronomy*, vol. 28, no. 3, pp. 471–484, 2008.

[AVI 07] AVIRON S., KINDLMANN P., BUREL F., "Conservation of butterfly populations in dynamic landscapes: the role of farming practices and landscape mosaic", *Ecological Modelling*, vol. 205, pp. 135–145, 2007.

[BAD 05] BADDELEY A., TURNER R., "Spatstat: an R package for analyzing spatial point patterns", *Journal of Statistical Software*, vol. 12, no. 6, pp. 1–42, 2005.

[BOU 10] BOUSSARD H., MARTEL G., VASSEUR C., "Layers dependencies specifications in the APILand simulation approach: an application to the coupling of a farm model and a carabid model", *2010 International Conference on Integrative Landscape Modelling – LandMod 2010*, Poster, Montpellier, France, 2010.

[BRO 09] BROGGI D., GenExp: Génération de paysages agricoles, mémoire de DUT en informatique, IUT Charlemagne, Nancy, June 2009.

[CAR 02] CARSJENS G.J., VAN DER KNAAP W., "Strategic land-use allocation: dealing with spatial relationships and fragmentation of agriculture", *Landscape and Urban Planning*, vol. 58, pp. 171–179, 2002.

[CAS 07] CASTELLAZZI M., PERRY J., COLBACH N., MONOD H., ADAMCZYK K., VIAUD V., CONRAD K., "New measures and tests of temporal and spatial pattern of crops in agricultural landscapes", *Agriculture, Ecosystems & Environment*, vol. 118, pp. 339–349, 2007.

[CAS 08] CASTELLAZZI M., WOOD G., BURGESS P., MORRIS J., CONRAD K., PERRY J., "A systematic representation of crop rotations", *Agricultural Systems*, vol. 97, pp. 26–33, 2008.

[COL 09a] COLBACH N., "How to model and simulate the effects of cropping systems on population dynamics and gene flow at the landscape level: example of oilseed rape volunteers and their role for co-existence of GM and non-GM crops", *Environmental Sciences and Pollution Research*, Springer, Berlin/Heidelberg, vol. 16, no. 3, pp. 348–360, 2009.

[COL 09b] COLBACH N., MONOD H., LAVIGNE C., "A simulation study of the medium-term effects of field patterns on cross-pollination rates in oilseed rape (Brassica napus L.)", *Ecological Modelling*, vol. 220, no. 5, pp. 662–672, 2009.

[DEL 04] DELAÎTRE J., Simulation de paysages agricoles aléatoires bidimensionnels, mémoire de DUT en informatique, IUT Charlemagne, Nancy, 2004.

[GAR 87] GARDNER R.H., MILNE B.T., TURNER M.G., O'NEILL R.V., "Neutral models for the analysis of broad-scale landscape pattern", *Landscape Ecology*, vol. 1, no. 1, pp. 19–28, 1987.

[GAR 91] GARDNER R., O'NEILL R., "Pattern, process, and predictability: the use of neutral models for landscape analysis", in Quantitative Methods in Landscape Ecology, TURNER M.G., GARONER R.H. (eds), pp. 289–307, Ecological Studies 82, Springer, New York, 1991.

[GAU 06a] GAUCHEREL C., FLEURY D., AUCLAIR D., DREYFUS P., "Neutral models for patchy landscapes", *Ecological Modelling*, vol. 197, no. 1–2, pp. 159–170, 2006.

[GAU 06b] GAUCHEREL C., GIBOIRE N., VIAUD V., HOUET T., BAUDRY J., BUREL F., "A domain-specific language for patchy landscape modelling: the Brittany agricultural mosaic as a case study", *Ecological Modelling*, vol. 194, no. 1–3, pp. 233–243, 2006.

[GAU 08] GAUCHEREL C., "Neutral models for polygonal landscapes with linear networks", *Ecological Modelling*, vol. 219, pp. 39–48, 2008.

[GOT 96] GOTWAY C.A., RUTHERFORD B.M., "The components of geostatistical simulation", *2nd International Symposium on Spatial Accuracy Assessment in Natural Resources and Environmental Sciences*, Fort Collins, USA, 1996.

[GUE 03] GUERREIRO A., Développement d'un générateur expérimental de paysages agricoles aléatoires bidimensionnels, mémoire de DESS, UHP Nancy 1, 2003.

[HAR 02] HARGROVE W.W., HOFFMAN F.M., SCHWARTZ P.M., "A fractal landscape realizer for generating synthetic maps", *Conservation Ecology*, vol. 6, no. 1, pp. 2, 2002. http://www.consecol.org/vol6/iss1/art2/

[KAL 07] KALUZNY L., KOCH O., Pavages du plan dans GenExp, mémoire de projet d'initiation à la recherche, master d'informatique, UHP Nancy 1, May 2007.

[KOS 10] KOSTRZEWA B., Réalisation d'algorithmes numériques pour un logiciel de simulation de cultures, mémoire de DUT en informatique, IUT Charlemagne, Nancy, 2010.

[KUR 00] KURZ W., BEUKAMA S., KLENNER W., GREENOUGH J., ROBINSON D., SHARPE A., WEBB T., "TELSA: the Tool for Exploratory Landscape Scenario Analyses", *Computers and Electronics in Agriculture*, vol. 27, pp. 227–242, 2000.

[LAV 08] LAVIGNE C., KLEIN E. K., MARI J.-F., LE BER F., ADAMCZYK K., MONOD H., ANGEVIN F., "How do genetically modified (GM) crops contribute to background levels of GM pollen in an agricultural landscape?", *Journal of Applied Ecology*, vol. 45, no. 4, pp. 1104–1113, 2008.

[LAZ 09] LAZRAK E.G., MARI J.-F., BENOÎT M., "Landscape regularity modelling for environmental challenges in agriculture", *Landscape Ecology*, Springer, vol. 25, no. 2, pp. 169–183, 1 September 2009.

[LEB 98] LE BER F., BENOÎT M., "Modelling the spatial organisation of land use in a farming territory. Example of a village in the Plateau Lorrain", *Agronomie: Agriculture and Environment*, vol. 18, pp. 101–113, 1998.

[LEB 06] LE BER F., BENOÎT M., SCHOTT C., MARI J.-F., MIGNOLET C., "Studying crop sequences with CARROTAGE, a HMM-based data mining software", *Ecological Modelling*, vol. 191, no. 1, pp. 170–185, 2006.

[LEB 09] LE BER F., LAVIGNE C., ADAMCZYK K., ANGEVIN F., COLBACH N., MARI J.-F., MONOD H., "Neutral modelling of agricultural landscapes by tessellation methods – application for gene flow simulation", *Ecological Modelling*, vol. 220, pp. 3536–3545, 2009.

[LIU 98] LIU J., ASHTON P. S., "FORMOSAIC: an individual-based spatially explicit model for simulating forest dynamics in landscape mosaics", *Ecological Modelling*, vol. 106, pp. 177–200, 1998.

[MAC 96] MACKISACK M., MILES R., "Homogeneous rectangular tessellations", *Advances in Applied Probability*, vol. 28, pp. 993–1013, 1996.

[MAR 06] MARI J.-F., LE BER F., "Temporal and spatial data mining with second-order Hidden Markov models", *Soft Computing – A Fusion of Foundations, Methodologies and Applications*, vol. 10, no. 5, pp. 406–414, 2006.

[MOR 06] MOREY M., TAYEBI T., Développement d'un logiciel de génération expérimentale de paysages (GenExpNextGen), mémoire de projet d'initiation à la recherche, master d'informatique, UHP Nancy 1, May 2006.

[OKA 00] OKABE A., BOOTS B., SUGIHARA K., NOK CHIU S., *Spatial Tessellations: Concepts and Applications of Voronoi Diagrams*, 2nd edition, John Wiley & Sons, Chichester, 2000.

[ORO 98] O'ROURKE J., *Computational Geometry in C*, 2nd edition, Cambridge University Press, Cambridge, 1998.

[SAU 00] SAURA S., MARTÍNEZ-MILLÁN J., "Landscape patterns simulation with a modified random clusters method", *Landscape Ecology*, vol. 15, pp. 661–678, 2000.

[SOR 08] SOREL L., Paysages virtuels et analyse de scénarios pour évaluer les impacts environnementaux des systèmes de production agricole, thèse de doctorat, Agrocampus Ouest, Rennes, 2008.

[TUR 91] TURNER M.G., GARDNER R.H., (eds), *Quantitative Methods in Landscape Ecology*, Ecological Studies 82, Springer, New York, 1991.

[WIT 97] WITH K.A., KING A.W., "The use and misuse of neutral landscape models in ecology", *Oikos*, vol. 79, no. 2, pp. 219–229, 1997.

Chapter 8

MDWEB: Cataloging and Locating Environmental Resources

8.1. Introduction

Environmental applications are on the rise, especially when it comes to territory development projects subject to stronger and stronger natural and human constraints (risk management, territory management, process observation and resource assessment, cultural heritage management, etc.). All these applications require geo-referenced data. Although these data are locally available and their volume is increasing, due to the explosion in the use of new data sources (GPS readings, airborne and satellite images, etc.) and tools to manage them (GIS, spatial database management system (DBMS) we must however go beyond the data produced to question their ability to be efficiently mobilized to serve the targets that are aimed for, whether we are going for monitoring, diagnosis, or decision helping. To this end, we must implement "decisional" systems allowing us to use the

---

Chapter written by Jean-Christophe DESCONNETS and Thérèse LIBOUREL.

data and processes based on multiple sources and focusing on a single space or a single issue. Designing such systems requires the means to gather necessary resources, make them accessible, and share them (if possible) in a unified manner.

A spatial data infrastructure (SDI) is one of the solutions which is suggested today to ensure the outcome. The architecture of these infrastructures always provides for a discovery service that relies on metadata specific to the shared resources [LIB 03]. It is an essential component that will help both resource dissemination and resource locating. One of the advantages of such a service is undoubtedly its ability to link the need to highlight these data inherent to the producers and the final users who are linked to the location and access to the resource.

This chapter presents MDWEB, an open source tool for cataloging and locating environmental resources (data and documents). This tool is an interesting example of complementarity between various actors, those from research (IRD, LIRMM, CEMAGREF), business (Geomatys), and from the social sphere (data users, managers, etc.).

8.2. Context

8.2.1. *Origins*

The MDWEB project was initiated in 2003. The design and development of the tool resulted from the functional analysis carried out within the frame of the Observatory Network of Long Term Ecological Monitoring ("Réseau d'Observatoires de Suivi Ecologique à Long Terme", ROSELT). The goal of this network was to understand and characterize the mechanisms of land degradation, its causes and consequences in circum-Saharan Africa [ROS 04a].

Studying these land degradation mechanisms in the 11 countries in question required mobilizing historical data to ensure the diachronic analysis essential to understanding long-term phenomena. It was thus essential to share the information due to the diagnostics of land degradation on each part of the land to compare the different situations encountered.

These requirements quickly created the need to inventory and describe the pre-existing, collected, and elaborated data. Another requirement was to respect the autonomy of the institutions involved in managing the data throughout the data collection, analysis, and interpretation phases of the monitoring. To this end, sharing the mutualization of these information sources quickly became very useful to build a common knowledge of the studied observatories [DES 03].

This institutional context and the issue of mutualizing data inherent to network-based work then led to the implementation of an information system based on a distributed architecture. This distributed architecture is the forerunner of the current SDIs [ROS 04b].

To provide one of the main components to this information system, we designed a tool ensuring a homogeneous description of the resources to be pooled and located: MDWEB. We structured it around metadata [DES 01]. The tool, in its first version, had four core functions: referencing, researching, locating, and accessing the data.

Functional evolutions followed, taking into account requests based on the projects in which we participated. We will specifically mention ACI PADOUE (inciting concerted action "Sharing data for environmental uses", *PArtage des données pour des Utilisations en Environnement*, 2002–2006), the SYSCOLAG program "Integrated Management of

COastal and LAGoon SYStems" carried out in the Languedoc-Roussillon region [MAZ 06], the PER project (rural excellence center, "pôle d'excellence rurale") [DES 07a] initiated by the DIACT (Interministerial delegation to land use and competitiveness), and the BibioMar project (2009) carried out by the DIREN (Regional Directorate of Environment) of Reunion Island.

Since 2007, MDWEB has undergone a technology transfer from Geomatys[1]. During this event, new technical orientations were decided [DES 08]. These were directed toward a component-based architecture, strongly service oriented (SOA for service-oriented architecture). These steps were taken with the goal to make MDWEB an essential and autonomous component of current SDIs such as the initiatives of the GMES European program (*Global Monitoring for Environment and Security*) and the INSPIRE European directive.

8.2.2. *Positioning*

SDIs are a solution to the rapid evolution of the Internet and standardization initiatives in terms of geographical information use, especially for organizations such as the Open Geospatial Consortium (OGC) and the International Organization for Standardization (ISO). Beyond the institutional, organizational, and legal aspects, from a computer science viewpoint, an infrastructure relies on a distributed architecture based on normalized services. These are the services that provide the primitive functions available to the user such as discover, location, visualization, and data download.

---

[1] http://www.geomatys.fr/

Discovery services are one of the ways to partially make up for distributed and heterogeneous resource interoperability through the descriptive role of metadata. Current search engines (Google, Yahoo, etc.) could potentially ensure such a discovery service, but metadata models on which they rely are unspecific, incomplete, and not explicit to the nature of the resources to be cataloged. That is the cost of their generic aspect. Moreover, they do not bring the flexibility (genericness) required to adapt such a service to a community's specific issue. Finally, indexing and search techniques are "proprietary" and thus depend on the engine.

Many normalization proposals relative to metadata have appeared in the past few years [DCM 05], SensorML [OGC 07a], Darwin Core [TDW 09], ABCD [TDW 06], in various fields such as digital documentation, geographical information, and biological information. When it comes to geographical information, following on the first proposals drawn up by the Federal Geographic Data Committee (FGDC, USA) at the end of the 1990s [FGD 98], the ISO 19115 standard, "Geographical information – metadata" [ISO 03] has been predominantly adopted and is recommended in most SDIs.

A tool combining resource cataloging through metadata models adapted to spatiotemporal resources with a search tool based on this metadata was in our view an essential addition to environmental applications.

In this context, resource cataloging initiatives are already widespread. The explosion of activities on the Internet confirms the interest of scientific communities to disseminate and share their resources and knowledge. There are a few solutions today that match this need to locate and access digital resources. Studying these software solutions [DES 07b] highlights that they are mainly implementing international standards (ISO, OGC) that shape catalog structure and ensure part of their interoperability. It also appears that the support chosen to deploy these tools is generally the Web, if we omit tools included in wider offers (ESRI's ARCCATALOG[2]) or designed a few years ago (the French Study Center on transportation networks and town planning REPORTS[3]).

Beside MDWEB, we will mention open source software, called *open source*, which exists in the geographical information community, GEONETWORK[4], developed by the Food and Agriculture Organization (FAO) of the United Nations, and its French version GÉOSOURCE[5], which was implemented by the Office for Geological and Mining Research (Bureau de recherches géologiques et minières – BRGM) to answer the needs for French cataloging due to the INSPIRE directives. These two tools are based on mainly equivalent architectures: a DBMS component ensuring the storing of metadata (ORACLE, MS-ACCESS or MYSQL, POSTGRESQL for the open source software) articulated around a Web application server such as TOMCAT/JAVA, APACHE/PHP, and IIS/ASP.

As for user interactivity, most of the tools today offer cartographic modules, which provide support to build geographically based queries and allow, for some, visualization of geo-referenced data. Metadata editing help is a function that is more or less developed depending on the tools.

8.3. Major functionalities and case uses

MDWEB offers two major functionalities. They are articulated around metadata management or "cataloging" and

---

2 http://www.esri.com/
3 http://www.certu.fr
4 http://geonetwork-opensource.org/
5 http://www.geosource.fr/

resource search to locate, choose, and access the desired resources, the "locating" function.

We will thus briefly remind the reader what we mean by cataloging (section 8.4) and locating (section 8.5) and add the administration functionality (section 8.6) that appeared necessary. All the roles accessing these functionalities will also be mentioned throughout the uses.

In the context of the data user, the tool must allow the search for a resource from metadata querying. This query is formulated by combining five criteria focusing on the content of the resource (what?), its type (what type of resource?), its spatial extension (where?), its temporal extension (when?), and the organization management or owning it (who?). The answer obtained will be selected by the user. After this selection, the user will potentially access the resource through a Web protocol (HTTP, FTP, etc.).

In the context of the data producer, the tool must ensure the description of the resources made available through this infrastructure. This description relies on metadata. Within the description function, we will separate the cataloging from the semantic annotation. The first can be considered as the technical description of the resources. The elements linked to cataloging will allow us to characterize the resources. The second notion, semantic annotation, expands the first to focus on the semantic description of the resource. It will rely on existing and expected metadata elements to describe the content of the resource.

8.3.1. *Matching roles and functionalities*

This tool is meant to be a component in a data infrastructure. We have set the perimeter of the infrastructure to be that of a limited user community (with a common field of interest). In this context, functionalities have been defined for the user of predetermined roles (actors in the UML sense). In the context of a multiuser tool, these different roles allow us to assign and organize the operations required to edit, validate, and publish metadata according to the level of expertise of the actors involved and their role in the organization in which the tool is used. Although the nomenclature is specific to MDWEB, the seven roles offered (see Figure 8.1) rely on functional segmentation inherent to multiuser tools.

The final user: he accesses the search module to locate relevant resources by viewing the afferent metadata, then accesses the resource if it is available online. The locating functionalities are publicly available, so this role is given to anybody in the community looking for resources, from the layman to the cartographer.

The commentator: linked to the production of metadata, he provides commentaries on the existing metadata files that will be useful to monitor editing and to validate metadata. This role is given to someone who has taken part in creating or has expertise over the described data. This can be, for instance, a technician who took part in gathering information.

The author: he inputs or imports metadata to describe and annotate the resource produced. This role is given to the person who is directly involved in creating the set of data, or the person who created it. It can either be the cartography engineer or the study engineer who took part in designing the data set specifications.

The validator: beyond the resource description work, he validates the meta-information of the producer before allowing its dissemination to the community. He makes sure of the quality of the metadata content. This role can be fulfilled by an expert in the field who can be a scientist, or a project manager.

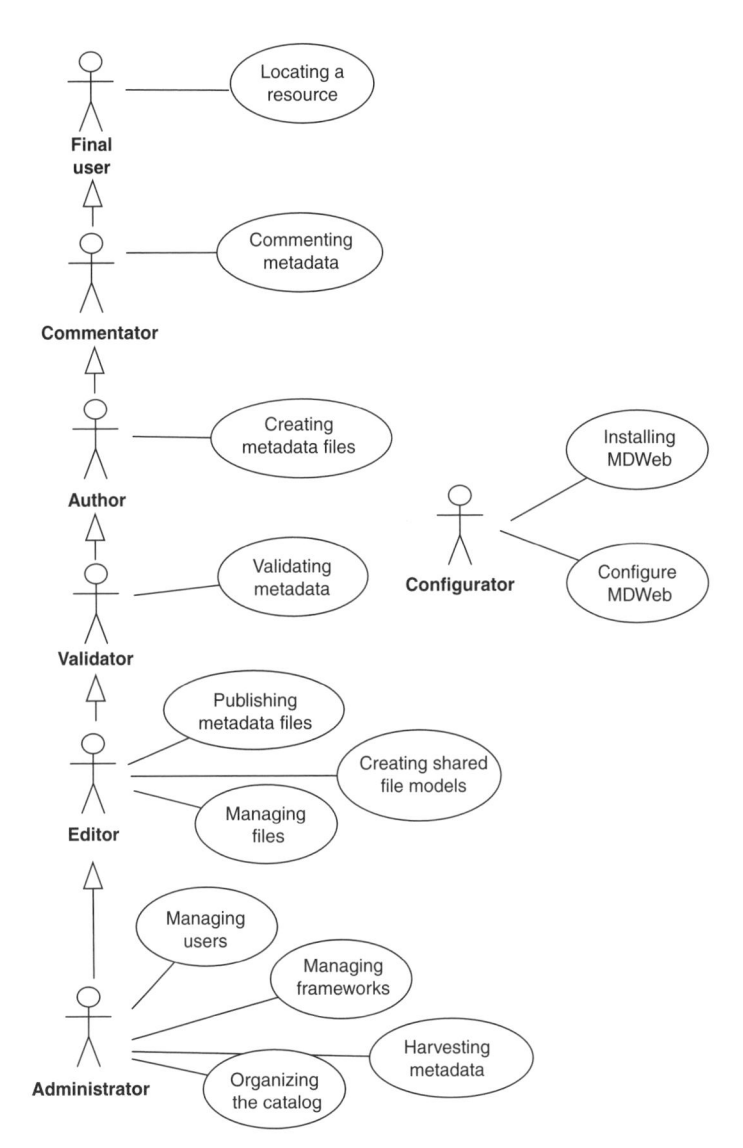

Figure 8.1. *Roles and matching functionalities (formalize use case UML)*

The editor: he publishes the validated metadata so that it can be queried by the search module. This role can be fulfilled by the field expert or the geomatics expert. A single person often endorses the roles of editor and administrator.

The administrator: he must define the structure of the meta-information (standard, metadata profile) and its organization within the tool (managing the sets of named files), choose the matching frameworks, and define the context of use within the tool (definition of the cartographic context, choice of thesauri). He is in charge of metadata management, and of the roles of the tool users. This role is given to the geomatics data administrator of the organization who has the required geomatics expertise and knowledge of the data processing tool.

The configurator: he installs the tool and configures it to ensure its proper functioning for other users. He does not take part in the editing, metadata administration, and tool administration processes. He is different from the administrator because we think that this role should be given to a system administrator who does not necessarily have knowledge in geomatics.

Each potential actor (besides the final user) is assigned an account and one of the roles defined above by the administrator. Depending on his role, he then has access to the functionalities matching the tasks he has been assigned.

8.4. Cataloging functionality

The cataloging functionality of the tool is based on metadata. Before we describe it in detail, we will present in sections 8.4.1 and 8.4.2 the notions of metadata, of metadata profiles that are widely used and contribute to the structure content of the catalog. Then we will detail the cataloging functionalities in a simplified manner (section 8.4.3) and in a multiuser context in section 8.4.4. Finally, we will provide in section 8.4.5 the extensions implemented in MDWEB to facilitate metadata management.

8.4.1. *Notion of metadata*

The notion of metadata is far from being new, but we must admit that today the use and importance of metadata have increased. This highlights the increasing needs for management and location of mass-produced information that is by nature heterogeneous, since it is the results of data production in formats and representations, dispersed with different producers.

In its first meaning, metadata means "data about other data, or data that provides information about other data and allows them to be relevantly used" [BER 93]. When there is no metadata, geo-referenced data in particular can be used only in a restrained manner and we cannot assess the quality of their content.

Thus, metadata must enable management, a relevant and wise use of data. It provides a potential user with the means to know their availability (format and access conditions), their accuracy in meeting specific needs, and the way they can access them (protocol). In our context, metadata elements on which the environmental resource cataloging relies today are imposed by existing standards. In general, metadata and standard structure obey models with more or less complex hierarchies [BAR 05]: "We can describe a metadata standard as an aggregation of sections, each made up of a set of structured metadata elements (...) linked to the description of a specific category of the information..." (see Figure 8.2).

Norm ◇—*— Section ◇—*— Metadata element

Figure 8.2. *Simplified view of the general organization of a standard*

Today, the ISO 19115 standard offers a wide and complete model of metadata for geographical information. It has become omnipresent in the environment community. This is a federating standard which is based on a synthesis of preexisting standards. Rather exhaustive, it offers a description of geo-referenced resources due to 12 sections (see Figure 8.3). The main sections cover the identification of the main characteristics of the resource (Identification), the information on access constraints (Constraints), information on resource quality, and information on maintenance and resource update. Finally, the ISO 19115 standard offers a section of metadata identification (Metadata) that brings together the elements of metadata file management (creation date, standard, standard version, metadata language, etc.).

A tool coupling the cataloging of resources through specific metadata models (adapted to spatiotemporal resources) with a search tool based on this metadata was, in our view, an essential contribution to environmental applications.

8.4.2. *Notion of metadata profile*

One of the points of the ISO 19115 standard is its flexibility to different communities, especially due to the implementation of profiles. This notion is offered in ISO standards (ISO 19106 and ISO 19115). A metadata profile can be considered to be a specialization of the standard. A user community can thus select mandatory elements from a metadata profile and add additional non-standard elements (for instance elements linked to the state of the sets of data produced). Figure 8.4 illustrates these two notions.

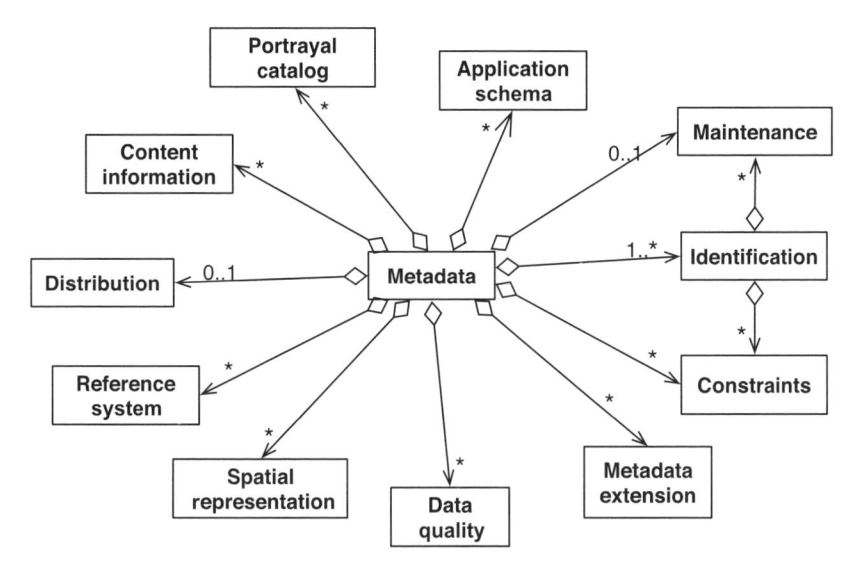

Figure 8.3. *Different sections for the standard ISO 19115 (UML formalism)*

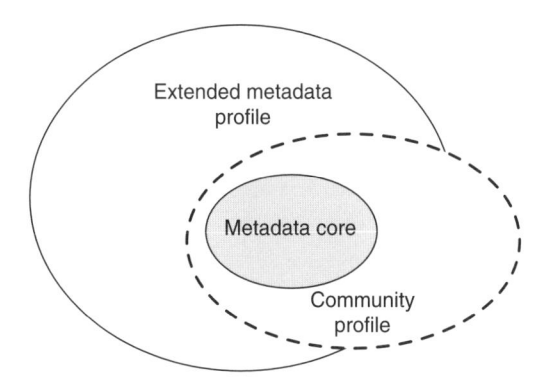

Figure 8.4. *Metadata profile*

We have completed this notion by linking a metadata type to a resource type. During the editing phase, we offer the user interfaces that filter and adapt the entry fields according to the type of resource to be described.

For example, when inputting a set of vector data, the profile used to generate the data entry form constrains the metadata element `SpatialRepresentationType` to the vector value, or offers the input of the class `MD_SpatialRepresentation` from ISO 19115 by using the `MD_VectorSpatialRepresentation` specialization that is devoted to the description of vector data, and contains elements describing the topology level (`topologyLevel`) or even the geometry type `geometricObjects`. Table 8.1 presents the types of resources and the matching metadata profiles offered by MDWEB by default.

8.4.3. *A simplified view of cataloging*

The cataloging organization in MDWEB revolves around the notion of metadata sheets. A metadata sheet relies on a profile and matches a set of values that provide information about the metadata elements in these profiles (see Figure 8.5). For instance, MDWEB offers a profile called "Vector layer", which is provided to describe geographical data in vector mode. This profile is adapted from the standard ISO 19115.

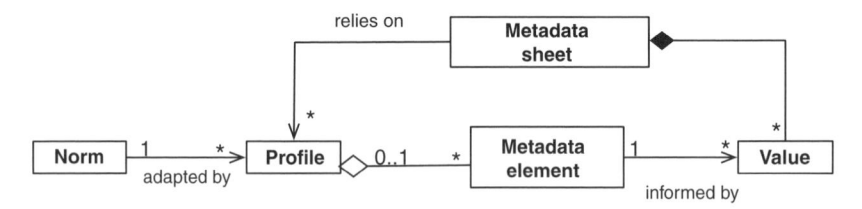

Figure 8.5. *The notion of metadata sheet (UML formalism)*

Thus, resource cataloging in MDWEB is mainly done by carrying out different applied management operations or operation sequences on a metadata sheet. These operations go from creation to publishing to annotation. The diagram of use case in Figure 8.6 shows the detailed view of the cataloging functionality of a resource in an organization where (to simplify) a single person is in charge of all the operations.

Type of resources	Definition	Known format examples	MDWEB profile
Digital document	Non-digital document, work document, internal report	DOC, PPT, XLS, OD*, PDF	ISO19115: doc_text_num
Paper document	Published document: work, extract of a work report, publications, scientific thesis	Hard copy or digital equivalent (see formats above)	ISO19115: refer_biblio
Digital map	Corresponds to storing, as a file, the result of the formatting of a set of information layers (vector, matrix) of graphical elements including a legend, a scale	mxd+ (ESRI), WOR+ (MAPINFO), mapset (GRASS), QGIS, JUMP, GVSIG, ORBISGIS	ISO19115: map_num
Paper map	Map created on a non-digital support	Paper document or digital copy	ISO19115: map_paper
Data table, spreadsheet	Raw data archived on a digital support as a table, with columns as the type of data, and lines the recordings	DB, CSV, TAB, ODT, XLS	ISO19115: table_num
Database	Relational database without geographical dimensions		ISO19115: base_alpha
Geographical database	Relational database, some of the attributes are geometries with a system of referenced coordinates	Geodatabase ESRI, set of GRASS maps, POSTGIS, ORACLE, MYSQL databases	ISO19115: base_geo
Raster layer	Matrix geographical data or *raster* mode	IMG, JPEG2000, ECW, TIFF	ISO19115: a_raster
Vector layer	Vector geographic data or vector mode	*shapefile*, MIF/MID, DWG	ISO19115: layer_vecteur

Table 8.1. *Resource typology and matching metadata profiles*

230   Innovative Software Development in GIS

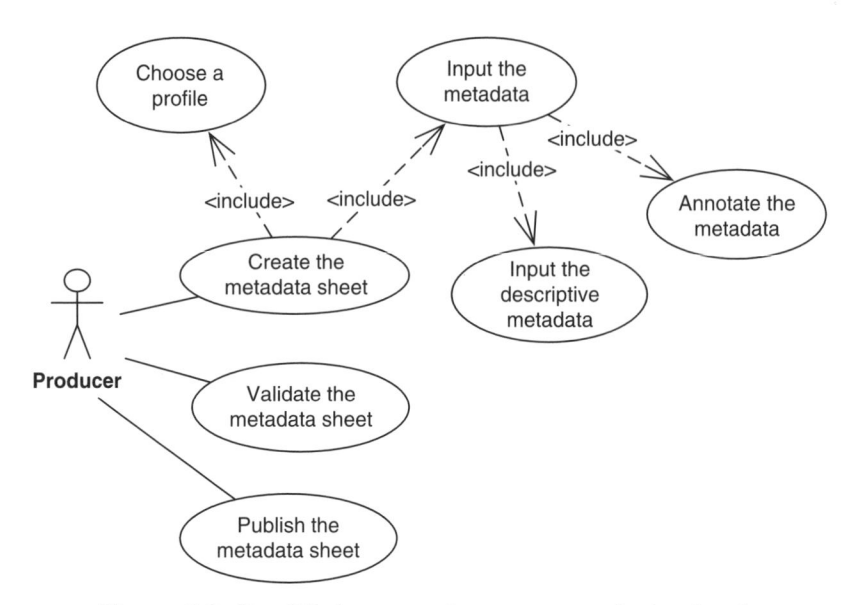

Figure 8.6. *Simplified case use for resource cataloging (UML)*

The first function is choosing a metadata profile matching the resource to be cataloged. The choice of this profile allows us to load the linked form from which the metadata will be input. During this operation, we have two sub-use cases: inputting descriptive metadata and semantic annotation of the resource through metadata.

The first sub-case covers inputting metadata elements that can be described as "descriptive". Indeed, these elements bring structure information about properties that are intrinsic to the resource, such as the main characteristics of identification, representation mode, spatial and temporal frameworks, the specification defining the data model, the description of the genealogy; or on information about distribution (access conditions, format, etc.), management and even information allowing us to identify the metadata itself (see Figure 8.6).

The second sub-case covers the input of metadata elements that can be called semantic. Indeed, these metadata elements included in the identification section, such as keywords, and topicCategory, are limited by a list of fixed values (enumeration). When it comes to the input of the element keywords, we have introduced the use of a thesaurus to control the terms on which the indexation and the research will rely. A thesaurus is a set of terms linked together by equivalence relations (synonymy), associative relations, or hierarchical relations [ZAY 10] defined by the standard ISO 2788 [ISO 86].

Various reference thesauri have been added to the MDWEB tool: the multilingual agricultural thesaurus AGROVOC of the FAO and the general multilingual thesaurus GEMET focusing on the environment of the European Environment Information and Observation Network (EIONET). Other than these thesauri, we have introduced the notion of thematic framework. A thematic framework describes, for a given community, the semantics of a field considered through explicit knowledge models. Any model translated the expertise of a community. It is the vector of semantic interoperability between actors to share understanding of the field's concepts. For the MDWEB tool, this framework is stored in a relation diagram allowing us to import models using semantic Web languages (SKOS, *Simple Knowledge Organization System*; RDF, *Resource Description Framework*).

A similar approach is implemented for information spatiality. We will call spatial framework a set of relevant geographical objects that are thus referents for the community in question. The community relies on the EX_Geographic-Extent section of the metadata ISO 19115 and provides the user with geographical objects of interest. These geographical objects can be represented in a various ways:

geographical names (terms of a thesaurus), encompassing rectangle describing the spatial extent, precise geometry (see Figure 8.7).

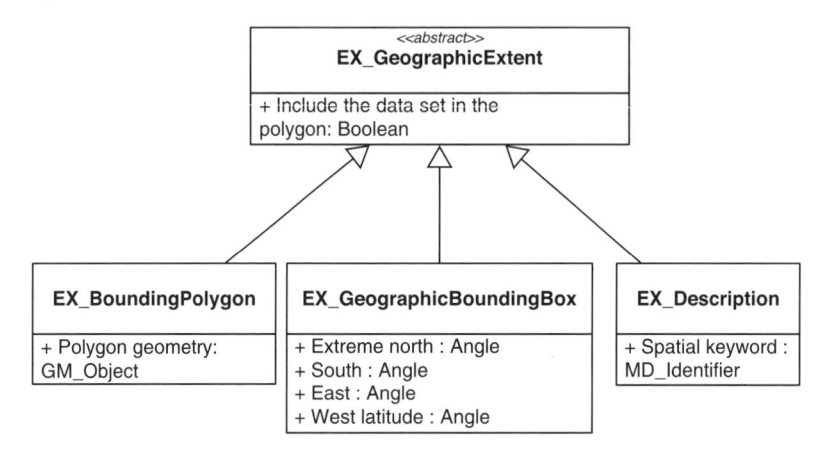

```
                    <<abstract>>
                  EX_GeographicExtent
          + Include the data set in the
          polygon: Boolean
```

EX_BoundingPolygon	EX_GeographicBoundingBox	EX_Description
+ Polygon geometry: GM_Object	+ Extreme north : Angle + South : Angle + East : Angle + West latitude : Angle	+ Spatial keyword : MD_Identifier

Figure 8.7. *Illustrating the three geographical describers of the standard ISO 19115*

The two thematic and spatial frameworks are closely linked. The spatial dimension is a mediator between thematic search and spatial search. A spatial term (of the thematic framework) describes the intent (for example, lagoon) of a concept, the latter has an extension in the shape of geographical objects that make up a layer of the spatial framework. Then, the data producer can carry out the validation of the input sheet himself and publish it to allow the search engine to query its content.

8.4.4. *Cataloging in a multiuser context*

The different functions linked to cataloging can be delegated to various users through roles that will be assigned to them in the tool (see Figure 8.1). This can be useful, and even necessary, within an organization that, due to its configuration (various disseminated teams) and the multidisciplinary aspect of its activities, must ensure the consistency of its metadata and control the publishing phase according to its charter. To this end, the different roles identified match the core competencies of the different levels that will be required for each operation.

In this context, the functions are shared along the role given to the user. Let us rework the use case diagram shown in Figure 8.6 to take this more complex vision of cataloging into account.

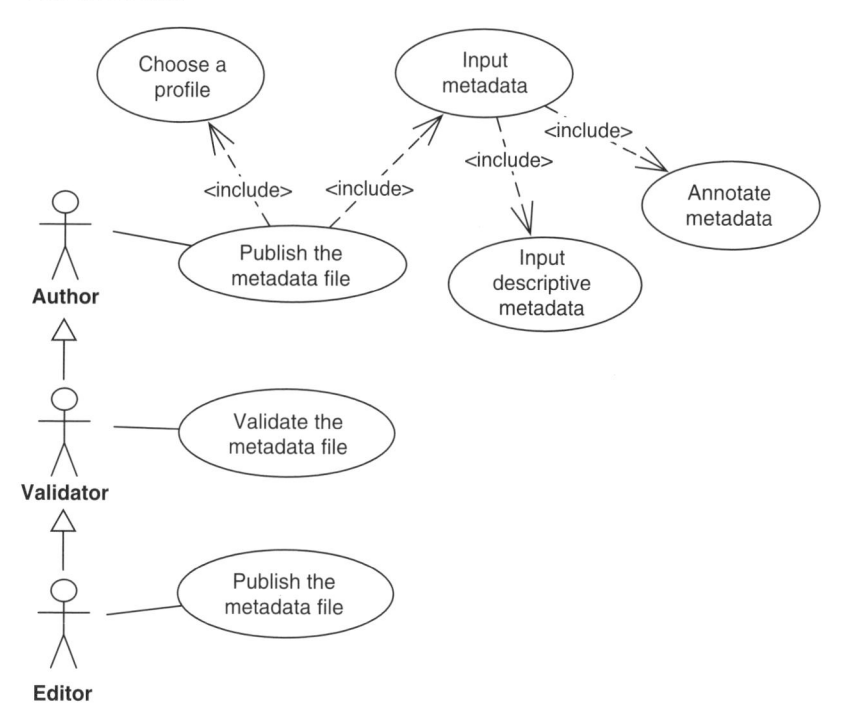

Figure 8.8. *Use case for (UML) resource cataloging*

Seeing this sharing of functions, managing a metadata sheet requires different actors to be involved. To better describe the various stages through which a metadata sheet will go, let us describe its lifecycle by using the formalism of UML activity diagrams. Figure 8.9 shows a global vision of the different actions and how they are linked, leading to the publication of a metadata sheet.

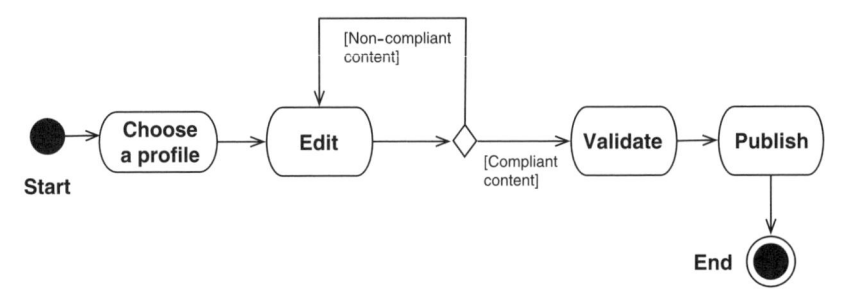

Figure 8.9. *Lifecycle of the publication of a metadata sheet (UML activity diagram)*

A sequence of four main actions is carried out:

– the choice of a metadata profile on which to rely when inputting a metadata import;

– editing that ensures the input/update of the metadata elements within the chosen profile, to describe the resource;

– the validation of the sheet's content or its compliance to the semantics and format of the metadata elements input; in case it is non-compliant, we will have to modify its content;

– publication within the tool.

8.4.5. *Cataloging extensions*

8.4.5.1. *Help for metadata input*

When editing metadata, the input phase can very quickly become a constraint, and, thus, lead users to lose interest. One of our core concerns was thus, when designing the tool, to help the user as best as we could in carrying out all the tasks they required. Therefore, MDWEB offers a set of facilitation tools.

The first tool helps edit a metadata sheet by prefilling all or more of the elements the sheet contains. To this end, we have implemented mechanisms that rely on the creation of sheet templates or sheet fragments. A sheet template can be defined as a metadata sheet in which certain fields were prefilled in order to be reused to create other sheets. This provides the author with a sheet template that he will be able to reuse to describe resources with identical properties such as distribution formats, contact point coordinates, and geographical extent.

The reuse principle can also be refined by the notion of sheet fragment. It will be used to load values within a specific section of metadata. We can then create fragments allowing us to fix the values of a section of metadata concerning the description of access constraints (`MD_Constraints`) or even contacts (`CI_ResponsibleParty`). The fragment is called up and loaded within the form when editing the sheet and will be reattached to the afferent section.

The second tool, within the multicriteria entry and search interfaces, relies on the semantic addition of the two frameworks mentioned in the previous section. The effort to help data entry focused on the elements used to index sheets (keyword section, spatial extent section). For example, inputting keywords is done by providing users with completion component based on the terms in the thesaurus (see Figure 8.10). A second level of use for this framework is possible when inputting data. To expand the field of accessible terms, a navigation interface ensures that we can browse the hierarchical relations and equivalences to the terms of the chosen thesaurus.

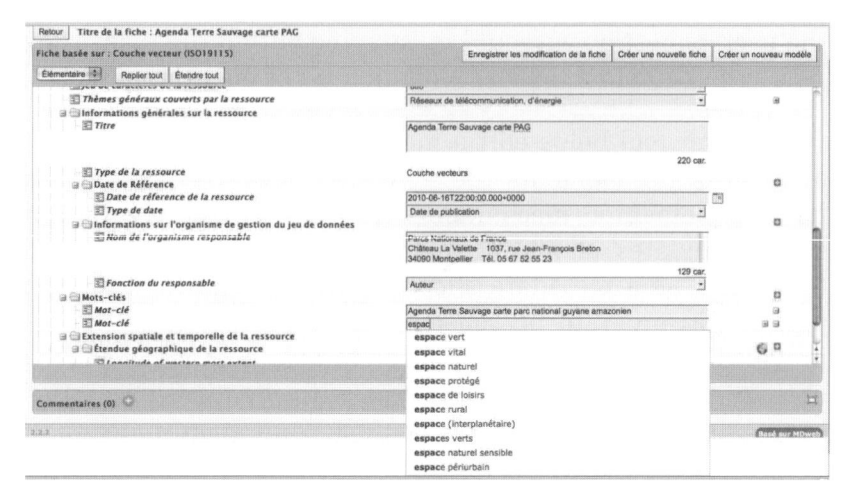

Figure 8.10. *Screenshot capture of the autocompletion component used during the editing phase to select the keywords coming from embedded thesauri*

The spatial framework exploitation is used to guide the user when filling in the spatial extension section by offering a cartographic interface representing the geographical objects of interest. The user can then either select an existing geometry and its geographical name, or define a specific spatial extent (see Figure 8.11). For a user in search phase, the same components can be used to define location criteria (geometry and geographical name, specific extent) and keywords.

8.4.5.2. *Metadata exchange*

The current environmental applications are made up of available heterogeneous systems in environments in which each producer provides his own metadata in his own catalog. Providing a global resource vision within a catalog requires gathering metadata produced by all the producers taking part in the data infrastructure.

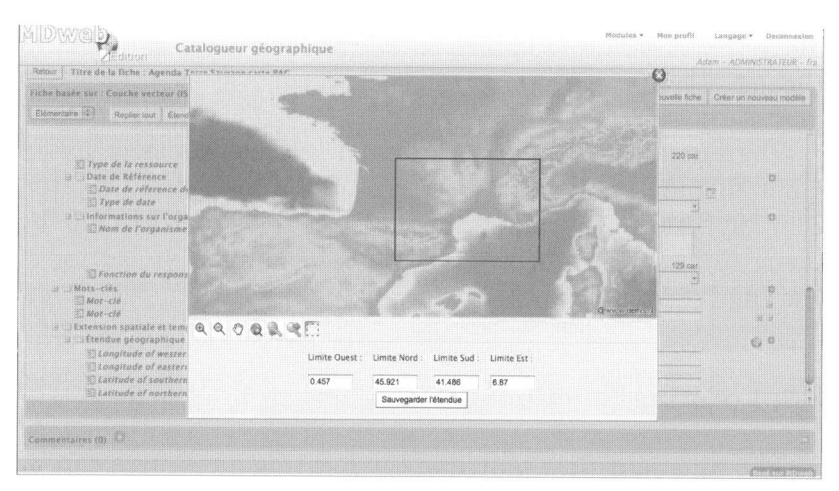

Figure 8.11. *Screenshot of the cartographic component used during the editing phase to help the user enter the spatial extension of the resource*

To this end, the cataloging functionalities are completed with metadata import and export function. Indeed, the metadata exchange standard is the ISO specification 19139 [ISO 07] that suggests an XML exchange format for metadata built with the ISO standard 19115. We use this format to transfer metadata. The author user can carry out a transfer during the editing phase. It can happen either by a set of files or file-by-file through a compressed file (archive) or an XML ISO 19139 file. Two scenarios are processed during import. In the case of the transfer of metadata created by MDWEB tools, the file's profile of the transferred metadata is identified and the import is complete. In the opposite case, when transferring from a different tool to MDWEB, the imported file's profile is not necessarily defined, and the user will have to define the matches by selecting an MDWEB metadata profile. Finally, the import process is carried out along two modes: a permissive mode that carries out the import and ignores the non-compliant elements and a strict mode that forces compliance on all the elements contained in the XML file to ensure import.

The automated import of a set of metadata sheets based on a distant catalog is also offered to the administrator. This operation is called harvesting. It is caried out by using the Harvest method of the cataloging Web service of the OGC.

8.5. Locating functionality

The functions that contribute to locating a resource can be presented by the use case diagram of Figure 8.12.

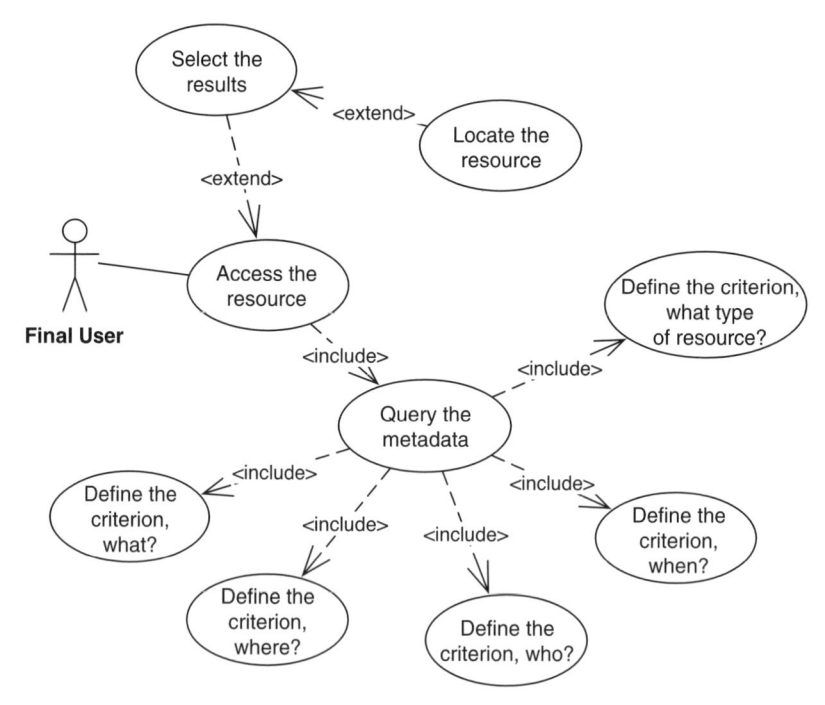

Figure 8.12. *Use case to locate (UML) resources*

First, the location covers metadata querying. MDWEB focused this query on the need to find a resource. By discovery, we mean the possibility of knowing the existence of a resource and being able to assess its content, and its main characteristics.

To efficiently get usable search results, we need to provide the user with the possibility to create queries of various complexity levels, ranging from a simple query involving the type of resource (what type of resource?), "I wish to locate vector type data" or "I wish to locate data created after September 3, 1990", to a query covering the nature and characteristics, especially the thematic (what?), spatial (where?), temporal (when?): "I wish to locate data about ZNIEFF protected spaces in the Languedoc–Roussillon region published during the nineties". We will explain the formulation of these queries by defining one to five criteria then combining them by the logical operator AND if necessary.

Let us precis the nature of each criterion.

– The "what?" criterion concerns the issue that is dealt with by the resource. Querying this criterion is based on various fields providing information about this content. This is the case of resource title (`title`), resource abstract (`abstract`), of the general topic(s) dealt with (`topicCategory`), and of the keyword field (`keyword`) for which we have introduced the use of a controlled vocabulary (see section 8.4).

– The "where?" criterion covers the geographical location of the resource of its spatial extension. This criterion can be answered by the values of the coordinates of a rectangle matching the spatial extent of the resource desired. Its input is facilitated by the availability of a cartographic interface offering geographical objects of interest. The coordinates of the encompassing rectangle will be accessible through interaction (a mouse click) with the selected object. This criterion matches the metadata elements linked to the description of the spatial extension of the resource contained in ISO 19115 class `EX_GeographicExtent`.

– The "when?" criterion covers both the period covered by the resource that can be assimilated to its temporal validity and the reference date of the resource. For the latter, it can be the date of creation, revision, or publishing. The criterion matches the metadata elements providing the temporal extension of the resource ISO 19115 `EX_TemporalExtent` and those matching the resource reference date, `date` and `dateType` within the identification section of the resource.

– The "who?" criterion concerns the organizations linked to the resource through their creation, management, or distribution. This criterion relies on the query of elements in the class ISO 19115 `CI_ResponsibleParty` and, more specifically, on those defining the name of the organization `OrganisationName` and its role when it comes to the resource, `role`.

– The "what type of resource?" criterion is to link with the resource typology offered in MDWEB (see Table 8.1). It allows us to filter the research by metadata profile.

These five criteria appear to cover most of the questions linked to discovering a resource. We will favor two of these criteria: what and where. They require geographical vocabulary and objects of the two frameworks embedded in the tool (see section 8.4). Thus, they allow us to make the metadata querying easier and more efficient.

The result of a query returns a set of metadata files matching the defined criteria. From this, the user can filter the results to analyze the characteristics of the desired resource in more detail. Other information will allow him to know the conditions and modes of access (Web service, ordering procedure, contacting the manager, etc.).

After this analysis, the user will select the resource(s) to visualize or download it (them), when the service is available online. The access conditions to the resource are described in the afferent metadata elements, classes ISO 19115 `MD_TransferOptions` and `MD_Constraints`. In the case of online resources, a Web address (`URL` or `URI`) is provided. It can point a static link toward the resource or a dynamic link toward a downloading service, visualization service, etc.

8.5.1. *Local and distant metadata querying*

By default, querying covers metadata stored in the current MDWEB tool. This is a local query. It can be extended to a querying covering metadata stored in distributed catalogs. This is a distant query.

MDWEB offers a unified query when we carry out a local or distant search. It relies on the implementation of the API offered by the OGC, the *Catalog Service for the Web* (CSW) [OGC 07b]. It provides methods to query metadata in one or more distant catalogs, and recommends languages to formulate the query and coding formats to retrieve the metadata.

Depending on the tool's installation context, either the addition of distant catalogs is the default, or it is carried out directly by the user. Sending the query will involve distant services whose `URL` was previously stored.

8.5.2. *Monolingual or multilingual querying*

Usually, MDWEB carries out a query based on the "what?" criterion by extracting the keyword of one of the thesauri in the language in which the metadata is informed. For example, if a user has a query on the keyword "espace protégé" (protected space), the results provided match the presence of the term "espace protégé" in the following metadata elements: title (`title`), abstract (`abstract`), keyword (`keyword`), and general topics (`topicCategory`). This is called a monolingual query.

In a transnational sharing infrastructure, this is not enough. Indeed, the query must be extended to formulate queries that will retrieve results no matter which language is used to describe the resources. This is called a multilingual query.

In both cases, the keywords used are extracted from thesauri linked to MDWEB. As we have previously mentioned (see section 8.4.3), the thesauri are structured by the resource description model RDF [W3C 04]. An RDF document is a set triple of subject, predicate, object. The subject represents the resource to be described; the predicate represents a property type applicable to this resource; and the object represents the value of the property that can be either data or another resource. An illustration is provided in Figure 8.13:

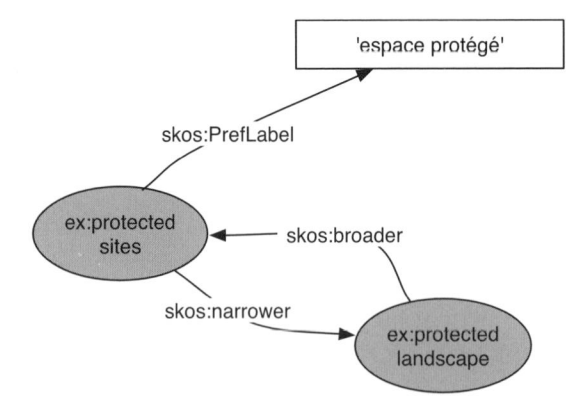

Figure 8.13. *Illustration of the RDF triple model with the* `protected sites` *concept*

The SKOS [SKO 04] vocabulary arises from the RDF model and is used to organize thesauri. In particular, it allows us to define a concept through its unique resource identifier (URI), which identifies it with no ambiguity. To this concept, SKOS allows us to attach a set of properties such as the label, prefLabel, the generalization relations, broader, and specialization relations, narrower, to other concepts. Thus, the concept protected sites is defined by the URI http://www.eionet.europa.eu/gemet/concept/6740 within the GEMET thesaurus whereas its PrefLabel property, whose value is "espaces protégés" in French (which is translated as "protected spaces"), is defined by the URI http://www.w3.org/2004/02/ skos/core#PrefLabel. By using SKOS syntax, this triple can be represented as follows:

```
<rdf:RDF>
 <skos:Concept rdf:about=
             "http://www.eionet.europa.eu/
                 gemet/concept/6740">
<skos:prefLabel xml:lang="fr">espaces
                 prot\'eg\'es</skos:prefLabel>
 </skos:Concept>
</rdf:RDF>
```

The PrefLabel property can be declined in other language to allow the concept to be matched to a Spanish term, for example, and the triple shown above thus becomes as follows ("espaces protégés" becoming "lugares protegidos"):

```
<rdf:RDF>
 <skos:Concept rdf:about=
             "http://www.eionet.europa.eu/
                 gemet/concept/6740">
<skos:prefLabel xml:lang="es">Lugares
                 protegidos</skos:prefLabel>
 </skos:Concept>
</rdf:RDF>
```

This organization allows us to extract various terms from the thesaurus by using the URI of the concept which they match. This mechanism is also indicated when we desire to carry out a multilingual query. Indeed, it will allow us to decline a keyword or expression, input by the user, in various desired languages. This declination relies on keyword translation. The notion of concepts offered by SKOS allows us to match the concept with the term in the desired language through its concept identifier (URI). The previous example is completed by providing the values of the prefLabel property of the *protected sites* concept in French, Spanish, and English.

```
<rdf:RDF>
<skos:Concept rdf:about=
              "http://www.eionet.europa.eu/
              gemet/concept/6740">
<skos:prefLabelxml:lang="es">Lugares
       protegidos</skos:prefLabel>
<skos:prefLabel xml:lang="fr">espaces
       protï¿½gï¿½s</skos:prefLabel>
<skos:prefLabel xml:lang="en">protected
       sites</skos:prefLabel>
</skos:Concept>
</rdf:RDF>
```

When solving the query about the term "espace protégé", the values in Spanish and English will be used to complete the query that is sent to distant cataloging services. MDWEB allows us to combine different query modes presented in these paragraphs to carry out, for instance, a local and multilingual query or a distant unilingual query.

8.6. Administration functionality

Within an SDI, MDWEB is installed in each organization in charge of data and metadata management. Its installation and configuration, and then the implementation and management of users and different functionalities, requires us to introduce an administration functionality linked to the first two functionalities described in sections 8.4 and 8.5. Figure 8.14 illustrates the main use cases linked to administration. There are two actors involved, the configurator and the administrator.

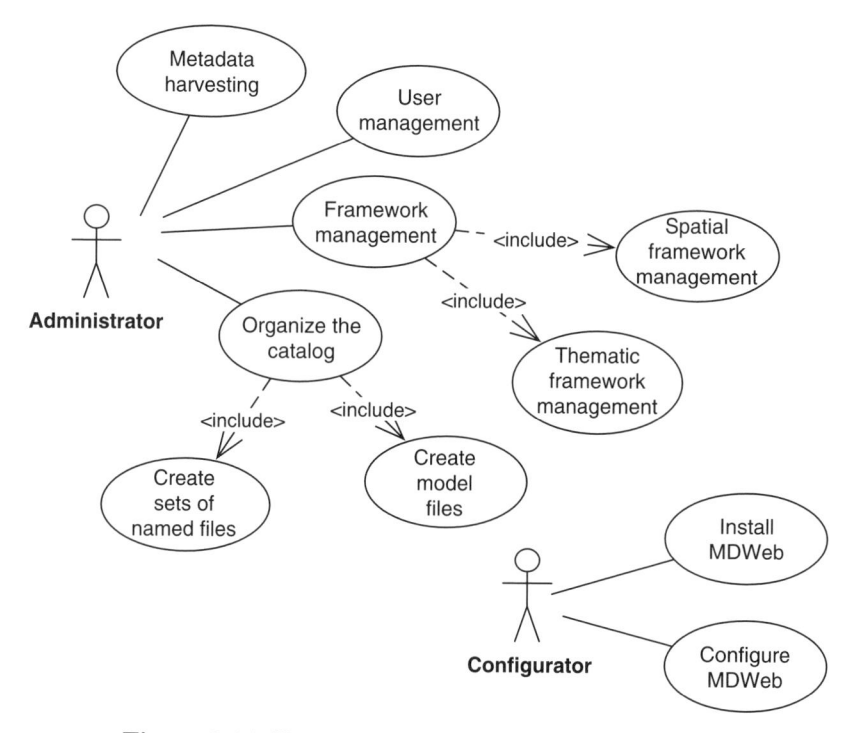

Figure 8.14. *Use case for the (UML) tool administration*

The configurator is linked to the functionalities that need only system competences. To this end, it is dissociated from the administrator that is considered as a field expert for which MDWEB is implemented. So the configurator deals with use cases that correspond to installing the tool and monitoring specified prerequisites for installation. It notably deals with implementing a JAVA application server and a DBMS.

Then, a configuration phase must be carried out step-by-step to create and people the database linked to MDWEB, parameter the connection to the database, and design the Web services used by MDWEB: the cartographic client of the research application (WMC, *Web Map Context*), the thesaurus Web service ensuring access to keywords for the research and editing applications, and finally the Cataloging service on which local and distant queries rely.

Only then does the administrator get involved to manage the components installed and the users that will be involved in metadata managements. To this effect, he is provided with four features:

– Framework management, which, in the current version, means choosing metadata profiles that will be available to authors for file editing, choosing the thesauri on which the semantic annotation and metadata queries will rely, and choosing the cartographic context used by the research application. In the latter, editing an XML file to the format given by the WMC specification of the OGC [OGC 05] provides the possibility of describing the geographical layers of the spatial framework and their representation style. This framework must be accessible through an OGC cartographic Web service (WMS, *Web Map Service*).

– User management means creating users and providing them with a role.

– Cataloging covers the creation of prefilled file models (matching the desired resource typology) and the creation of a set of named files. The file models created by the administrator are made available to the community for their future use. The sets of named files (such as "Fauna" and "Flora") are used to structure the catalog internally. They can be seen as folders/directories in which files about a certain subject or thematic or given organization are stored.

– Metadata harvesting, mentioned in section 8.4.5, refers to the automatic import of metadata from a distant catalog.

8.7. Architecture

For reasons of adaptability, modularity, and interoperability, MDWEB's architecture is based on components and is an SOA. Indeed, MDWEB must be able to interact (interoperate) with other tools or components relying on metadata within a data infrastructure. In the same way, it must call upon existing services to complete its components. From a logical point of view, according to the n-tier principle, MDWEB is articulated around three levels: the management level consisting of the metadata base and thematic framework, the dialogue level for data-HTTPD server and DBMS, and the application level (JSF, JAVA *Server Faces*). The tool is developed to be used independently on operating systems such as Windows, MacOS, and Linux. MDWEB is a software suite including client modules and a Web service server. It is all hosted and managed by a JAVA JEE application server: GLASSFISH 3 or APACHE TOMCAT 6. This software architecture ensures MDWEB's great flexibility when deploying, especially to allow it a unique deploying (MDWEB client modules and service server within a same server) up to a deploying based on a distributed architecture made up of multiple application servers and Web service servers.

Figure 8.15 depicts the main elements of this architecture. The client modules are accessible through a Web navigator. There are four and they cover the functionalities described in the previous sections:

– data research through metadata matching the locating functionalities shown in section 8.5;

– editing metadata that corresponds to the cataloging and annotating functionalities described in section 8.4;

– administrating catalogs and users, including the creation of named sets of recordings;

– designing the tool offering configuration functionalities for the tool and matching frameworks described in section 8.6.

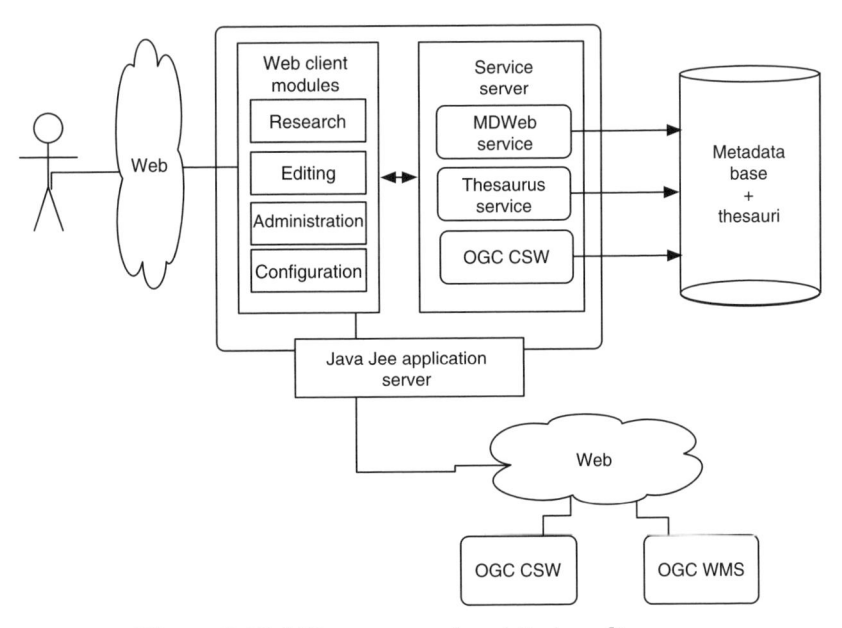

Figure 8.15. MDWEB *general architecture diagram*

These modules are developed with the JAVA Server Faces *framework*. This framework was chosen because it is component oriented and thus favors functional stability of developments. This is essential in a long-term project such as MDWEB. The MDWEB service server provides the following elements:

– A standardized CSW of the OGC providing research interfaces and access to the metadata files stored by MDWEB.

It is completed by the transactional mode (CSW-T) to ensure harvesting operations on the metadaa files coming from distant catalog.

– A specific MDWEB Web service prodiving interfaces to edit metadata, and administrate and configure the tool.

– A Web service to access the thesauri, enabling us to use the concepts and relations stored in these thesauri. This service is called upon during research or editing operations. It is based on the GEMET[6] Web service API.

– A relational database ensuring metadata, thesaurus, and used metadata models (standards and profiles) persistence.

The MDWEB server mainly relies on the geographic software component kit CONSTELLATION[7] developed by the Geomatys. It provides the SOA and software components that implement the geographic services of the OGC used within different modules of MDWEB. CONSTELLATION notably relies on GEOAPI[8], which offers the set of JAVA interfaces compliant with OGC specifications, and on GEOTOOLKIT[9], which implements part of these interfaces.

8.8. User community

Since its publication, MDWEB has been used in various fields and by various communities. Most often, it is used:

– for data inventory or administration reasons in an organization:

---

6 https://svn.eionet.europa.eu/projects/Zope/wiki/GEMETWebServiceAPI
7 http://www.constellation-sdi.org/
8 http://www.geoapi.org/
9 http://www.geotoolkit.org/

- new technology center and risk management, http://pont-entente.org/,

- national center for research support[10];

– to promote informational legacy:

- CIRAD cartographic atlas[11],

- rural excellence center[12];

– to implement an observatory:

- Banc d'Arguin National Parc observatory[13],

- regional program of maritime and coastal zone preservation in West Africa[14];

– to contribute to sharing data within a management project integrated to a territory:

- SYSCOLAG[15] program;

– as a discovery service within an SDI:

- *Best Practice Network for SDI in Nature Conservation*[16].

Since 2005, the MDWEB project has had its own Website[17] and a set of tools ensuring the monitoring of code releases[18],

---

10 http://cnar.mdweb-project.org/
11 http://cataloguecirad.teledetection.fr/
12 http://diact-demo.teledetection.fr
13 http://pnba.mdweb-project.org/
14 http://soga.univ-brest.fr/mdweb/
15 http://syscolag.teledetection.fr/
16 http://www.naturesdi.mdweb-project.org/french-geoportal/search/main.jsf
17 http://www.mdweb-project.org
18 http://hg.mdweb-project.org/mdweb/

of bug fix requests[19], a continuous integration platform[20], and user forum[21].

The Web site also offers the old and current release of downloads and the matching material to ease installation, and use of the tool. MDWEB is released under two open source licenses. Its PHP version (version 1.X) is disseminated under a CeCILL[22] open source license compatible with the GPL license. Version 2 is released under LGPL 3.0[23]. By analyzing downloads on the site, we can venture that MDWEB is widely used in the French and Francophone communities (France, Belgium, Canada, and French-speaking Africa). The tool has been downloaded more than 10,000 times (for all versions).

Engineers and computer science students contributed to version 1.X. There were three study engineers (for 5 years) and 10 computer science masters interns from the University of Montpellier II for these different versions. The Research Institute for Development (IRD) initiated the program within its Desertification unit, and later its Space unit. There was a strong collaboration that sprung up then with the D'OC team of the LIRMM, and it is still ongoing today. Other research organizations in Montpellier, such as the CEMAGREF (UMR TETIS), the CIRAD, and the CEPRALMAR, participated in a more occasional manner.

8.9. Conclusion

We wanted to offer a complete tool based on open source technology that could be modular and adaptable, especially

---

19 http://jira.codehaus.org/browse/MDWEB
20 http://hudson.geomatys.com/job/MDweb2/
21 http://mdweb-project.869954.n3.nabble.com/
22 http://www.cecill.info/
23 http://www.gnu.org/licenses/lgpl.html

when it comes to making it available and personalizing the frameworks matching the targeted application field. These are some of the points on which MDWEB is different from other open source cataloging solutions from the start. Moreover, it appeared essential to build a tool that was both generic and extensible, according to the standard ISO 19115 and the OGC specifications, using the frameworks matching the targeted field.

The generic approach was validated many times over the period, both by the diverse use made of the tool and by its adaptability to communities' needs. To this end, we have been able to implement other metadata models than ISO 19115 without compromising the architecture and storing diagrams of the metadata. One such example is the Dublin Core or Sensor ML (SML) metadata models which are the suggested models according to the OGC specifications, covering access to observations from Web sensors (SWE, *Sensor Web Enablement*).

Although metadata is often badly perceived, current projects and initiative of environmental information sharing, especially through the implementation of SDIs, allowed us to promote the use of metadata among the environmental communities. More often than not, these initiatives start out by creating their metadata service before deploying the other services of the infrastructure. That said, there are still limits that can be an impediment to the role given to metadata. Indeed, formulated and implemented metadata within discovery services are often created at the beginning of the project and then suffer from real and permanent updates. This then creates the issue of the freshness of the information provided. From a descriptive point of view, this type of metadata is still limited today to the discovery of resources and does not offer a correct description of the genealogy that would allow us to assess accurately their adequacy versus targeted thematic, temporal, or spatial precision. An effort in this direction should be made to better assess and describe the quality of resources aimed at the targeted uses in diagnostic or natural risk management fields.

8.10. Bibliography

[BAR 05] BARDE J., Mutualisation de données et de connaissances pour la gestion intégrée des zones côtières. Application au programme SYSCOLAG, PhD Thesis, University of Montpellier II, 2005.

[BER 93] BERGERON M., Vocabulaire de la géomatique, Office de la langue française du Québec, Montreal, 1993.

[BER 01] BERNERS-LEE T., HENDLER J., LASSILA O., "The semantic web: a new form of web content that is meaningful to computers will unleash a revolution of new possibilities", *Scientific American*, vol. 1, pp. 34–43, 2001.

[DCM 05] Dublin Core metadata initiative, DCMI Abstract Model, 2005.

[DES 01] DESCONNETS J.-C., LIBOUREL T., MAUREL P., MIRALLES A., PASSOUANT M., "Proposition de structuration des métadonnées en géosciences. Spécificité de la communauté scientifique", *Géomatique et espace rural, Journées Cassini 2001*, Montpellier, France, 2001.

[DES 03] DESCONNETS J.-C., MOYROUD N., LIBOUREL T., "Méthodologie de mise en place d'observatoires virtuels via les métadonnées", *Inforsid 2003, Actes du XXIème congrès*, Nancy, France, 2003.

[DES 07a] DESCONNETS J.-C., LIBOUREL ROUGE T., CLERC S., "Cataloguer pour diffuser les ressources environnementales", *Inforsid 2007, Actes du XXVème congrès*, 2007.

[DES 07b] DESCONNETS J.-C., MAUREL P., VALETTE E., LIBOUREL T., TONNEAU J., "Excellence et innovation rurales. Outil Web de gestion des données et référentiel d'analyse de projets PER pour un développement territorial durable", *Congrès joint du 47e ERSA (European Regional Science Association) et du 44e ASRDLF (Association de science régionale de langue francaise)*, Paris, France, 2007.

[DES 08] DESCONNETS J.-C., HEURTEAUX V., "MDweb 2.0: a java/JEE metadata catalog", *FOSS4G 2008, Free and Open Source Software for Geospatial Conference*, Cape Town, South Africa, 2008.

[FGD 98] FEDERAL GEOGRAPHIC DATA COMMITTEE, IContent standard for digital geospatial metadata – FGDC-STD-001-199, 1998.

[ISO 86] INTERNATIONAL STANDARD ORGANIZATION, ISO 2788 – Guidelines for the establishment and development of monolingual thesauri, 1986.

[ISO 03] INTERNATIONAL STANDARD ORGANIZATION, ISO/TC211, ISO 19115 – Geographic Information Metadata, 2003.

[ISO 07] INTERNATIONAL STANDARD ORGANIZATION, ISO/TC211, TS/ISO 19139 – Geographic Information – Metadata – XML schema implementation, 2007.

[LIB 03] LIBOUREL T., DESCONNETS J.-C., MAUREL P., MOYROUD N., PASSOUANT M., "Les métadonnées: pour quoi faire?", *Géo-événement' 2003*, Paris, France, 2003.

[MAZ 06] MAZOUNI N., LOUBERSAC L., REY-VALETTE H., LIBOUREL T., MAUREL P., DESCONNETS J.-C., "A transdisciplinary and multi-stakeholder approach towards integrated coastal area management. An experiment in Languedoc-Roussillon", *Vie et Milieu* (Life & Environment), vol. 4, pp. 265–274, 2006.

[OGC 05] OPEN GEOSPATIAL CONSORTIUM, Web Map Context Documents, 2005.

[OGC 07a] OPEN GEOSPATIAL CONSORTIUM, OpenGIS Catalogue Services Specification 2.0.2 – ISO Metadata Application Profile, 2007.

[OGC 07b] OPEN GEOSPATIAL CONSORTIUM, Sensor Model Language (SensorML) Implementation Specification – OpenGIS Implementation Standard, 2007.

[ROS 04a] ROSELT/OSS, Organisation, fonctionnement et méthodes de ROSELT/OSS, Collection ROSELT/OSS, document scientifique, Observatoire du Sahara et du Sahel, 2004.

[ROS 04b] ROSELT/OSS, Système de circulation de l'information ROSELT: Définitions des métadonnées et élaboration des catalogues de référence, Collection ROSELT/OSS, contribution technique, Observatoire du Sahara et du Sahel, 2004.

[SKO 04] SKOS Simple Knowledge Organization System, 2004.

[TDW 06] TAXONOMIC DATABASE WORKING GROUP, ABCD, Access to Biological Collection Data (ABCD) Primer, 2006.

[TDW 09] TAXONOMIC DATABASE WORKING GROUP, Darwin Core terms, a quick reference guide, 2009.

[W3C 04] WORLD WIDE WEB CONSORTIUM, RDF, Resource Description Framework. W3C Recommendation, 2004.

[ZAY 10] ZAYRIT K., Modèles de données adaptés à la construction partagée d'un thésaurus dédié aux traits fonctionnels, mémoire de stage de Master 2, University of Montpellier II, 2010.

Chapter 9

WebGen: Web Services to Share Cartographic Generalization Tools

9.1. Introduction

This chapter describes WebGen, a framework offering cartographic generalization operations as Web processing services. Originally, WebGen was designed with the purpose of being a common research platform allowing people to share generalization algorithms, and contribute to more and more advanced experiments. Very soon, however, it became apparent that the architecture might also have benefits for other types of users, such as national mapping agencies (NMAs) and GIS providers. While the original version of WebGen used the Soap protocol for service descriptions and communication mode, the following release was compatible with the Open Geospatial Consortium (OGC)'s *Web Processing Services* (WPS), which are widely supported by GIS. The latest version is available as *open source*.

Chapter written by Moritz Neun, Nicolas Regnauld and Robert Weibel.

9.2. Historical background

Cartographic generalization is the process that enables us to create a map readable at a given scale from often highly detailed geographical data (such as the data acquired by national mapping agencies). These data can only be represented as such at a very large scale. As soon as the scale is reduced, the objects on the map become too small to be read by the reader. Figure 9.1 shows this phenomenon in more detail. A map at a scale of 1:10 k quickly becomes unreadable if it is reduced. However, the map at a 1:50 k scale has a simplified content: the objects represented on it have been exaggerated and thus the map can be scaled down. Generalization is thus the process allowing the cartographer to simplify the map content to obtain a good compromise between the degree of detail and readability. There are other factors than scale involved in the choice of the objects to be exaggerated, aggregated, or eliminated. They depend on the needs of the target map user.

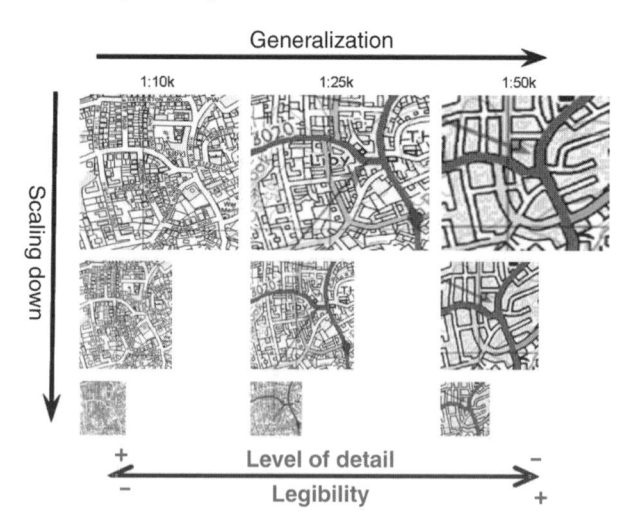

Figure 9.1. *Scale reduction and generalization*

The thematic content is dictated by the intended map purpose. The actual representation of content then depends not only on the scale at which the map must be displayed or printed, but also on the type of medium used to display the map, the purported map reader, and the reading context (day or night, for instance). The generalization process has an important cognitive aspect. The goal is not so much to create a map complying with strict specifications as much as to create a map that contains the information the user needs so he/she can interpret the map accurately.

Map generalization for a long time was a completely manual process. Its automation today represents a major challenge for national mapping agencies. It would help reduce map production costs considerably and open the door to on-demand map creation, generated for specific purposes. Research into the automation of this process has thus been carried out for more than 20 years (since the arrival of digital geographical data). The cognitive aspect linked to cartographic generalization, however, makes automation extremely demanding. Various solutions were suggested for different aspects of the process, using a great variety of often complex techniques. The International Cartographic Association (ICA) Commission on Generalization and Multiple Representation brings together a community of researchers and industrial actors working on the subject. This community published a book [MAC 07] that gathered together a series of articles covering the main aspects of the issues of automatic generalization. [RUA 02] offers a similar compendium in French.

Faced with such a complex problem, the research community realized in the past few years that it needed a common development platform. This would help come up with more complex and complete solutions, reusing what had already been accomplished, focusing on new aspects. A major part of the research in map generalization has been carried out within the context of academic doctoral dissertations, which invariably have a fixed duration. If these researchers have to develop their prototypes from scratch, they reach a certain ceiling which is hard to pass. Conversely, if they can take up the development made by their predecessors, the ceiling disappears. This is a slightly simplistic view of the matter, but it clearly illustrates the benefits which such a platform would bring to the research field.

For a few years now, there has been much progress in the fields of interoperability, distributed architectures, and especially in the standardization of formats for geographical information transfer, mostly thanks to the actions carried out by the OGC. This led a group of researchers to write the first article [EDW 03] as the technologies became available to build it. This study was presented during a workshop of the ICA Commission on Generalization and Multiple Representation in 2003. The subject has since then been presented and debated in all the workshops of this commission. Work started at the University of Zurich to build a distributed platform using Web service technology. WEBGEN, a prototype framework to deploy generalization operators as Web services, was thus born. WEBGEN, described in [NEU 05, NEU 08], enables us to embed automatic processes written in JAVA, for instance, in methods presenting a generic interface, to post them on a server, and to use them from a remote desktop. The platform allows the user to find available services and invoke them by providing them with control parameters as well as the data to be processed. The processing is carried out server side and the results are sent to the client. Additional testing was subsequently carried out at the research department of Ordnance Survey (the British national cartographic agency) to assess the interoperability of these services with a specific platform. We performed the tests with RADIUS CLARITY, a platform devoted to developing generalization processes, developed, and marketed by 1Spatial[1]. The test consisted in checking that services could be called from RADIUS CLARITY, and also that processes native to the platform could be published as services, and therefore made available to other plateforms. It was demonstrated that it is relatively easy to request WEBGEN services from RADIUS CLARITY and vice versa to publish an algorithm depending on RADIUS CLARITY, and request it in the *open source* GIS JUMP[2]. The results sparked the interest of part of the community and a workshop was organized in November 2007 in Southampton (at Ordnance Survey) to decide on the following steps to be taken to promote the use of such a platform. The GIS vendors present at this meeting requested that this platform be supported by an acknowledged standard.

The main argument was that these companies often wish to implement OGC standards. A platform built along such standards could thus be easily coupled to these GIS without requiring costly additional developments. The prospect of having a platform that could be used with the majority of GIS led to the decision to redevelop WEBGEN as an extension of the OGC's WPS standard. A more technical workshop took place in January 2008 in Zurich to define the new platform's specifications [FOE 10]. An implementation carried out at the University of Zurich and funded by Ordnance Survey then created a WEBGEN server along the new specifications, as well as a client for JUMP. Documentation was also produced to help adapting these clients and servers to proprietary platforms. The sources and documentation are available on the Web site of the ICA Commission on Generalization and Multiple Representation[3].

---

1 http://www.1spatial.com/software/sware.php?id=8
2 http://www.vividsolutions.com/JUMP/
3 http://aci.ign.fr/web_service.php

The platform is now ready for use. A first exploitation phase is now required to understand how to use it and how to improve its functionalities.

9.3. Major functionalities

WEBGEN allows us to publish software tools devoted to generalization as Web services based on the OGC's WPS standard. Available functions are described in the following sections.

9.3.1. *Uploading software tools*

A WEBGEN server allows a developer to publish his/her software generalization tools online as Web services based on the OGC's WPS standard. There are two main constraints: the first is that the tool must read the data it needs and pass the result on in a format recognized by the WEBGEN server. In the WPS standard, *vector* data are transferred between the client and the server in the GML format. The current WEBGEN prototype can either pass this GML format data directly on to the algorithm or convert it into *shapefile* or into a native JAVA format (e.g. JAVA Topology Suite for JUMP). The second constraint is that the tool does not depend on a proprietary platform or on its internal functionalities (such as 1Spatial's RADIUS CLARITY or ESRI's ARCOBJECTS). This latter constraint can be removed by adapting the server to make it work with a specific proprietary platform. An experiment carried out at Ordnance Survey (GB) showed that it was possible to publish WEBGEN services depending on the RADIUS CLARITY platform.

The experiment went as follows. When a WEBGEN service was invoked on the server, it launched an executable file which carried out the following operations:

– declare the environment variables required by the RADIUS CLARITY platform;

– load the function libraries of the platform;

– create an internal database in the platform, in our case, a database in the Gothic format;

– translate the data passed to the service, in our case, translate GML data into the Gothic format;

– load the objects into the database;

– translate the required function parameters: in the test, it was a function creating a *buffer* around a given geometry; the only parameter was thus an integer representing the *buffer* width;

– call the function: in the test, it was the buffer_create function, but we could create any type of function based on the function libraries in Gothic and call it in the same way;

– translate the results from the Gothic format to the GML format to create the result sent back by the service.

The adaptation thus mainly concerned the translation of data from the format used by the WEBGEN server (JAVA Features, GML or *shapefile*) and the internal format used by the proprietary platform, and it had to be ensured that the server could execute the tool on this platform. It also had to be ensured that the license allowed this type of proprietary platform use.

9.3.2. *Requesting a service*

The tools published on the WEBGEN platform can be used by anyone who chooses one of the following access modes:

– By looking up the service description and building his own http requests. This use is obviously the least satisfactory since it is completely manual and subject to many errors.

– By using a programming client that facilitates the design of requests and the reception of responses. This client usually is a function library in JAVA, C, or any language used by the client platform.

– By using a graphical client which is a *plugin* of a GIS platform. For now, there is only a single WEBGEN client available developed for the *open source* GIS JUMP. It is possible to adapt the client to other platforms and there is documentation to facilitate this process. A client prototype for 1Spatial's RADIUS CLARITY platform has actually been successfully developed at *Ordnance Survey* to test the ease with which the graphical client could be adapted to different platforms. Just like the server adaptation described above, the main modification is in the translation of the data from the GML format used by the standard WPS to the internal format used by the GIS platform.

9.3.3. *Cataloging and discovering services*

As part of the WPS standard, each server can provide on-demand the list of services it offers through the `GetCapabilities` request. However, a user might want to access services offered by different servers. Thus, the implementation of the WPS standard in the WEBGEN server was extended with its own service registry. In this registry, an organization can add a list of WPS servers with the services it is interested in and which are located elsewhere. The `GetAllCapabilities` request to this registry will provide the list of all the available services on its own server as well as on the other servers listed (see Figure 9.2). The idea behind this registry can in part be compared to the UDDI standard used with Web services based on the SOAP technology. However, since the OGC does not currently offer such a service registry for WPS, we have opted for this solution.

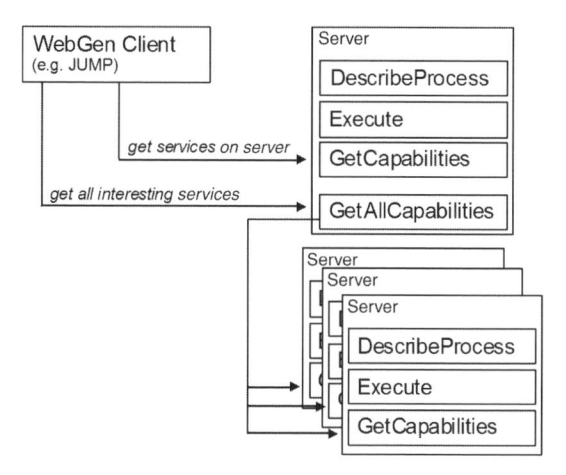

Figure 9.2. *Registry use as a* `GetAllCapabilities` *request*

9.4. Area of use

Different use modes have been envisioned for the platform to answer the needs of different types of users.

9.4.1. *Usage*

The platform was designed to allow two modes of use: the first is an interactive mode, useful to test and assess available services; the second is an automatic mode, which allows us to use available services to integrate them in more complex processes.

9.4.1.1. *Interactive mode*

The interactive mode allows the user to choose the service he/she would like to invoke, to select the data on which the service must be applied, and to input the parameter values requested by the service. To use this mode, a client with a graphical interface is required. The client is often an extension specific to a given platform (often a GIS which allows the visualization and manipulation of geographical data). The client accesses a register listing the servers offering WEBGEN services. It queries these servers to build a list of available services. Once the user has picked the service he/she wishes to use, the client reads the description of the service to learn about the expected parameters and builds an interface to allow the user to specify their values. The client also allows the user to choose the data (geographical objects) on which the service has to be applied. It must then translate these data in the format used by the platform into the standard format supported by WEBGEN. Once the service has returned a result, the client must translate it back into the data format supported by the platform.

Figure 9.3 shows the interface proposed by a WEBGEN client developed for the *open source* JUMP[4] platform. It allows the user to specify the address of the server to be accessed. The list of available services on this server is then displayed (left side of the figure). When the user selects a specific service, the parameters required by the service are displayed in the right-hand side of the window. In this example, the BufferFeature service requires three input parameters: width (an integer), capstyle (an enumeration), and quadrants (an integer).

9.4.1.2. *Automatic mode*

The automatic mode allows the services to be accessed from a program. This enables the creation of complex processes by chaining requests to existing services. Such a process can itself be published as a service. This allows us to consider a hierarchy of services. Building a complete generalization system did not just require generalization operations, it also

---

4 http://www.vividsolutions.com/JUMP/

required spatial analysis functions. These functions help the global generalization process to identify geographical structures implicitly present in the data. These functions enrich the data by building an explicit representation of these structures which can then be used to guide the generalization process. We thus offer a hierarchy of three WEBGEN service layers:

– The first layer, called support layer, essentially contains the lower level spatial analysis tools.

– The second layer is made up of generalization operators. They are elementary generalization operators (simplification, aggregation, etc.) for which various classifications have been offered, see for instance [REG 07c]. They are especially useful to enrich data by measuring the intrinsic characteristics of geographical objects or by qualifying their relations (distances, similarities, etc.) with their neighborhood.

– The third layer is made up of more complete processes which combine generalization operators to carry out more complete tasks, going as far as the complete generalization of a dataset.

This hierarchy makes the reuse of existing software tools easier, which helps the development of more and more complete and efficient tools. Figure 9.4 summarizes these three levels of services and their potential uses (interactive or automatic).

9.4.2. *User types*

We have identified three types of use for this platform.

9.4.2.1. *Researchers*

Researchers in automatic generalization (of which a great number are members of the ICA automatic generalization and

268   Innovative Software Development in GIS

multiple representation commission) should benefit from this platform in various ways:

– assess existing tools and carry out *benchmarking*;

– reuse exiting tools and concentrate only on developing more advanced new tools;

– carry out research on how to engineer generalization functions by using existing and accessible services in a standardized manner;

– carry out research into distributed and parallel processing, to improve performance (*cloud computing*);

– obtain feedback (comments, assessments) on the tools developed and published;

– promote the researcher's tools and thus their work and expertise.

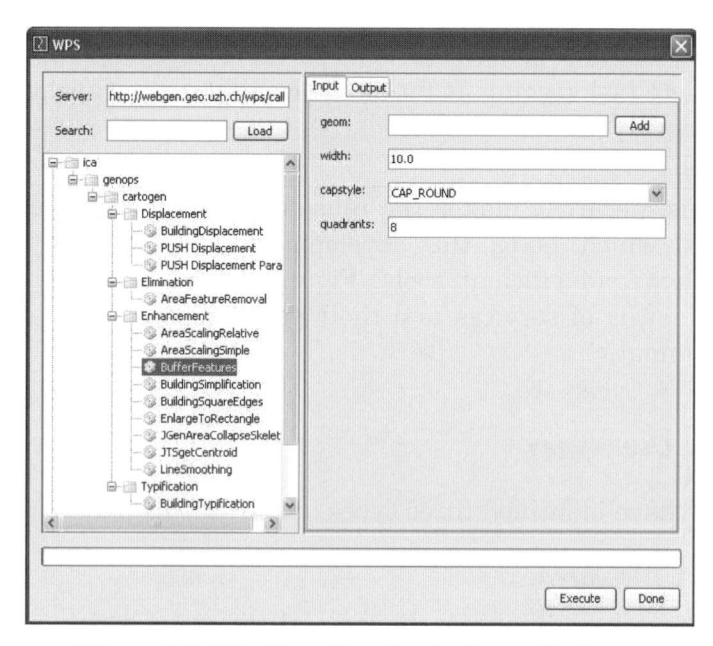

Figure 9.3. WEBGEN *client for the* JUMP *platform*

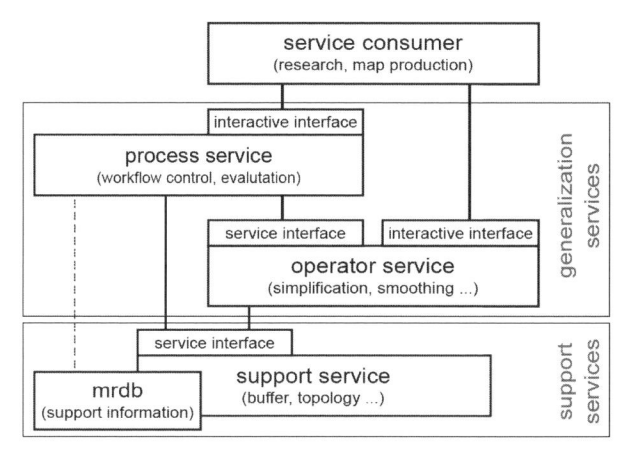

Figure 9.4. *Service hierarchy*

There are already a few uses of the WEBGEN platform in research.

[NEU 08] experimented with the development of a support layer (the "support services" in Figure 9.3) to help develop more complex services. The creation of complex generalization processes (*process service*) based on the engineering of existing operators is presented in [NEU 09]. In this context, parallel processing techniques can be used to accelerate a process.

The parallel processing of generalization tasks requires either the domain or a function to break down [LAN 91]. Breaking down the domain means cutting up a task into independent units. For instance, the data can be partitioned into disjointed zones, which can be processed independently (by partitioning the space using a road network, for example). The functional breaking down divides a process into tasks, some of them have to be processed sequentially and some of them can be carried out simultaneously.

[NEU 09] offers a multiple instruction and multiple data stream approach based on the breaking down of data

270   Innovative Software Development in GIS

and instructions into independent tasks to process them in a parallel manner. The method ensures that there is no concurrent data access issue. This approach can use a system with multiple processors, or a cluster made of various systems connected by a nework (*Cloud computing*).

Both the domain and functional breaking down can be used in a Web service environment using a cluster of generalization service instances, as shown in Figure 9.5. The *parallel processing service* plays the role of master process, partitioning buildings, and creating an instance of the *processing service* (1) for each partition. Each instance of the service is an independent process, sending data and receiving results. For each instance of the *processing service*, a *processing strategy* is carried out. The latter is a mechanism that allows us to explore a set of potential solutions (such as *hill climbing*). For each cycle, all the available operators are applied in a parallel manner to the partition's current state (2). The results are then assessed with the help of special services assessing whether certain constraints have been satisfied. This assessment of the different results also happens in parallel (3). Once all the assessments are done, the best solution is kept and the *process strategy* goes on to the next cycle. The parallel computing thus happens at three different levels of the processing.

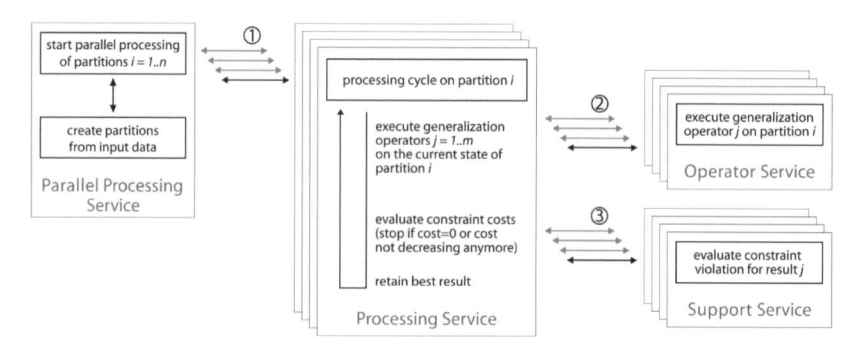

Figure 9.5. *Parallel service requests*

9.4.2.2. *Cartographic institutions (Institut Géographique National - IGN and others)*

Cartographic institutes could take advantage of the platform in two different contexts:

– to design new production systems:

  - to asses exiting tools by testing them on their own data;

  - to test existing tool sequences to create and assess production system prototypes;

– to develop a production system:

  - using such an architecture to develop a production system allows the use of tools working in different environments. This offers great flexibility for combining the best tools available for each task without being held back by limited availability of tools in a specific system.

Along these lines, [REG 07a] and [REG 07b] propose a system architecture able to automatically derive, through generalization, on-demand maps. This system is based on the use of Web services to access in a standardized manner a rich library of generalization tools from various platforms. This idea was developed after a conclusive experiments on the integration of WEBGEN and RADIUS CLARITY, the platform Ordnance Survey uses for its research on automatic generalization.

9.4.2.3. *GIS providers*

GIS providers could use the platform in various ways:

– assess the tools developed in research to choose which one to integrate in their systems. The fact they can test the tools provides them with a higher level of confidence and lowers the costs/risks linked with tool redeveloping;

– identify the collaborators to develop their systems;

– assess/compare their own tools versus other tools published as services;

– carry out a virtual extension of their systems by developing an interface which allows the Web services to be called up from their systems;

– provide generalization function as a paying service (*pay per use, cloud computing*). This would also allow small map designers to use efficient advanced production systems;

– promote their own systems or some of their tools by publishing targeted demonstrations as services.

Some GIS providers have already turned toward the concept of Web services to develop their software. Inspired by the success of the initial WEBGEN research project, a joint research project between the University of Zurich and Axes System was established to link the AXPAND software to external generalization functionalities by using Web services. The article [PET 06] shows how generalization sequences can be built by chaining generalization Web services. [BOB 08] discusses the use of the Web service concept to build and maintain links between geographical objects. Finally, a test with the University of Hanover has shown how easy it was for their commercial software PUSH, which performs automatic displacement of cartographic objects [SES 99], to be published as a WEBGEN service. A trial version of this service, limited to 250 objects, is available online[5].

This platform should allow all users to promote existing tools, reuse them, assess them, and thus have a better understanding of their strengths and weaknesses. It should also help collaboration between different types of users.

---

5 http://webgen.geo.uzh.ch/wps/

9.5. Architecture

WEBGEN is a platform built on a client–server architecture (see Figure 9.6). Its goal is to bring together in a virtual toolbox various tools useful to develop automatic generalization processes. It allows its users to access these tools in a standardized manner from various platforms, through the Web.

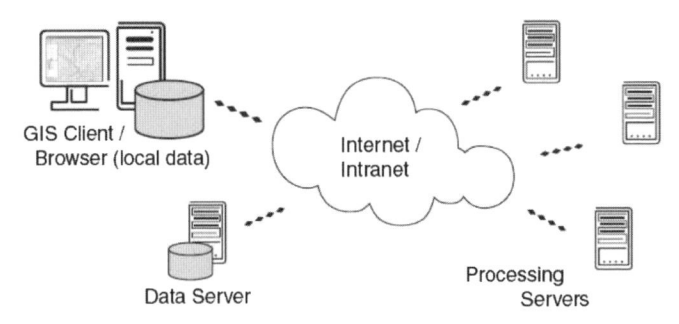

Figure 9.6. *Client–server architecture*

9.5.1. WEBGEN *services access*

WEBGEN is based on the OGC's WPS standard. WPS services are not the usual kind of Web service (like WMS, WFS) since they do not offer spatial data but processing capabilities, such as spatial algorithms. A WPS service can process data provided by the client ("upload") or data accessible on a server (like a WFS server or a *shapefile* on a Web server). To use WPS services (and thus WEBGEN), a client has access to three functions:

– getGapabilities: this method is implemented by the WPS server and provides the list of services offered by the server, with a short description.

– describeProcess: this function is implemented by each service and provides a detailed description of the service

274   Innovative Software Development in GIS

as an XML format document, notably including the required parameters and their type.

– execute: this method is implemented by each service and causes the execution of the service.

Figure 9.7 illustrates a typical exchange between a client and a server. The client starts by requesting the available services from the server. It then asks a specific service for a description of its parameters, and the data format required on input and output. It can then send the execution command to the desired service, with all the necessary parameters. Finally, the client gets the result of the execution in the predefined format.

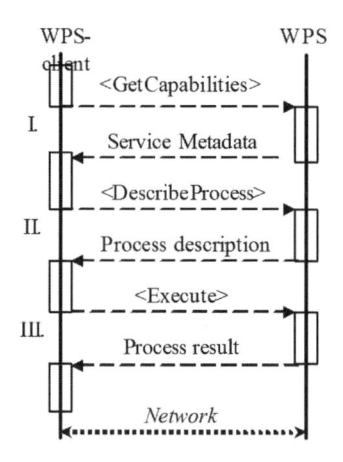

Figure 9.7. *Client and WPS server exchanges*

9.5.2. *A standard data model for generalization services*

The WPS service interface specifications offer great freedom in terms of data and parameters. The specifications do not impose any restrictions on the input or output data format. To facilitate the interoperability of these services, an extension like WEBGEN must restrain the formats used.

It was first decided to only use GML to encode the data at input and output of the WEBGEN services. The question of offering a standard to describe structures specific to the generalization processes (for instance, graphs) was also raised. Using a standard model would allow us to standardize the service description and to make their automatic selection possible (automatic discovery of an appropriate service). An architecture for on-demand map creation was suggested in [REG 07a]. This architecture partly relies on the availability of generalization tools as services, described in a standardized manner.

To use WEBGEN as a fully-fledged research platform, we must define a set of data formats and data models to improve the service interoperability as well as the compatibility between services and different clients. Moreover, we should plan on adding more complex data structures, such as graphs [NEU 08] and cartographic constraints [NEU 09] in the future. The classification proposed by [FOE 08] to extend the WPS standard provides a starting point. The following list provides an insight into the type of data which should be available on the research platform:

– literal data types (string, integer, floating point, boolean, date, and time);

– complex data types:

- geometries (encoded in GML 2): point, line, polygon, multipoint, multiline, multipolygons, and geometry collections;

- *features* (encoded in GML 2): collections of attributes (literal or complex), metadata (optional);

- collections of *features* (encoded in GML 2): collections of *features* with the same pattern, attributes (optional), and metadata (optional);

- list: list of literals or complex types;

- *map*: list of key-value pairs;

– more complex types to come: constraints [NEU 09], graphs [NEU 08], symbolization, and complex objects.

9.6. Associated communities

9.6.1. *Distribution*

The WEBGEN platform is mainly written in JAVA. The source code and literature are available as open source, licensed under GPL[6].

9.6.2. *Uses*

WEBGEN has been used for research purposes at the University of Zurich and has been tested at Ordnance Survey to experiment on the integration of different generalization tools coming from different platforms in a single production system. Recently, tests have been carried out in the GIS Department of the University of Lund (Sweden). These are for now the only identified cases of the use of WEBGEN.

9.6.3. *Contributors*

WEBGEN's original code was written by Moritz Neun at the University of Zurich during his doctorate. Moritz Wittensöldner ported WEBGEN toward OGC's WPS standard and wrote the documentation linked to it.

Various contributors took part in implementing the generalization services: Dirk Burghardt, Patrick Lüscher,

---

6 http://aci.ign.fr/web_service.php

Moritz Neun, Stefan Steiniger (all from the University of Zurich); Nicolas Regnauld (at *Ordnance Survey*); Stathis Perikleous (at *Ordnance Survey*, to publish an algorithm originally developed at the University of Edinburgh).

9.7. Conclusion and outlook

WEBGEN was designed to be a platform for generalization researchers, answering a need voiced by a large group of experts [EDW 03]. Since its original version, described in [NEU 05], the software was extended to be compatible with OGC's WPS standard and is now available as open source. Various studies [NEU 05, NEU 08, NEU 09] have shown that WEBGEN, a platform based on Web services for generalization, met the expectations and provided the originally targeted users (i.e. the researchers) the expected benefits. Moreover, there are certain national mapping agencies [REG 07a, REG 07b] and some cartographic software providers [PET 06, BOB 08] who found the service-oriented architecture concept attractive to build a generalization system. All the studies and projects have shown that the service-oriented architecture implemented in WEBGEN allows users to upload their own tools easily so that others can use them, and conversely allows them to easily use tools developed by other parties. In spite of the incredibly encouraging results reached in the three targeted user fields, WEBGEN is still not as widespread as the group of researchers who had originally suggested the idea thought it might be [EDW 03]. One possible explanation is the existence of other solutions also using a service-oriented architecture. For instance, the approach presented in [FOE 10] uses a combination of WPS and XML pattern translation to build automatic generalization processes. We believe, however, that XML pattern translation places considerable impediments on the sophistication and performance of the generalization tools which can be implemented.

Another (theoretical) reason for WEBGEN's lack of market penetration might be that generalization researchers are not as interested as they initially said they were in the creation of a common and sustainable generalization tool development platform. In today's academic system, researchers are rather rewarded for short-term innovations and for publishing algorithms in journals, rather than undergoing the long-term effort of creating robust algorithms and distributing them to other researchers in an easy-to-use manner. As long as there is no (financial) incentive, researchers might accept making their "raw" source code available, however, with no documentation attached, and in various formats (often specific to a particular development platform); but they will rarely make the effort of publishing their algorithm as a service. This final stage requires an extra effort to develop the interface between the algorithm and the standard format used by the Web services. One solution could be to find extra funding to convert the code and write the documentation. However, given current priorities in academic research, this seems unlikely. Today, a more realistic solution would be that the members of the ICA Commission on Generalization and Multiple Representation sign a memorandum of understanding on the development of service-oriented generalization tools. [FOE 08] lays out the principle behind this approach.

Although the factors presented above have affected the distribution of WEBGEN, there are other factors which are controlled by the platform contributors. First of all, the use of the platform in its current state should be better promoted. Until now, the publicity effort was minimal. For instance, placing the software on a portal like `SourceForge.net` would certainly increase its visibility.

The WEBGEN services would also be easier to use if a standard data model existed to describe the services (see above). Such a data model would not only solve the issue of the syntactic description of the service interface (what parameters are required and how they must be represented), but also provide information on the capabilities and conditions of use of the service (what does the service do, what are the typical fields of use, and what are the limitations). Finally, the best way to promote WEBGEN is probably to keep on showing the value of the service-oriented approach for cartographic generalization through new studies. This could, for instance, include a continuation of the studies initiated by [NEU 05] and [NEU 09]: research on the use of the platform to dynamically chain services, within the field of on-demand map creation, as well as research on parallelization in order to accelerate the execution of complex processes.

9.8. Acknowledgments

The original version of WEBGEN was developed at the University of Zurich by Moritz Neun while he was working on his doctorate, funded by the Swiss National Science Foundation (SNF) through the DEGEN project (SNF no. 200020-101798). The extension to the original prototype, to make it compatible with the OGC's WPS standard, was funded by the Ordnance Survey (GB). We wish to express our gratitude toward the organizations which took part in the funding as well as toward the colleagues mentioned above who contributed their ideas and algorithms.

9.9. Bibliography

[BOB 08] BOBZIEN M., BURGHARDT D., NEUN M., WEIBEL R., "Multi-representation databases with explicitly modeled horizontal, vertical and update relations", *Cartography and Geographic Information Science*, vol. 35, no. 1, pp. 3–16, 2008.

[EDW 03] EDWARDES A., BURGHARDT D., BOBZIEN M., HARRIE L., LETHO L., REICHENBACHER T., SESTER M., WEIBEL R., "Map generalisation technology: addressing the need for a common research platform", *21st International Cartographic Conference*, Durban, South Africa, 2003.

[FOE 08] FOERSTER T., BURGHARDT D., NEUN M., REGNAULD N., SWAN J., WEIBEL R., "Towards an interoperable web generalisation services framework – current work in progress", *Proceedings of the 11th ICA Workshop on Generalisation and Multiple Representation*, Montpellier, France, 2008.

[FOE 10] FOERSTER T., LEHTO L., SARJAKOSKI T., SARJAKOSKI L.T., STOTER J., "Map generalization and schema transformation of geospatial data combined in a web service context", *Computers, Environment and Urban Systems*, vol. 34, no. 1, pp. 79–88, 2010.

[LAN 91] LANGRAN G., "Generalization and parallel computation", *Map Generalization: Making Rules for Knowledge Representation*, Longman, London, pp. 204–216, 1991.

[MAC 07] MACKANESS W.A., RUAS A., SARJAKOSKI L.T. (eds), *Generalisation of Geographic Information: Cartographic Modelling and Applications*, International Cartographic Association, Elsevier, 2007.

[NEU 05] NEUN M., BURGHARDT D., "Web services for an open generalisation research platform", *8th ICA Workshop on Generalisation and Multiple Representation*, A Coruña, Spain, 2005.

[NEU 08] NEUN M., BURGHARDT D., WEIBEL R., "Web service approaches for providing enriched data structures to generalisation operators", *International Journal of Geographical Information Science*, vol. 22, no. 2, pp. 133–165, 2008.

[NEU 09] NEUN M., BURGHARDT D., WEIBEL R., "Automated processing for map generalization with web service", *GeoInformatica*, vol. 13, no. 4, pp. 425–452, 2009.

[PET 06] PETZOLD I., BURGHARDT D., BOBZIEN M., "Workflow management and generalisation services", *9th ICA Workshop on Generalisation and Multiple Representation*, Portland, OR, USA, 2006.

[REG 07a] REGNAULD N., "A distributed system architecture to provide on-demand mapping", *Proceedings of the 23rd International Cartographic Conference*, Moscow, Russia, 2007.

[REG 07b] REGNAULD N., "Evolving from automating existing map production systems to producing maps on demand automatically", *10th ICA Workshop on Generalisation and Multiple Representation*, Moscow, Russia, 2007.

[REG 07c] REGNAULD N., MCMASTER R., "A synoptic view of generalisation operators", *Generalisation of Geographic Information: Cartographic Modelling and Applications*, Elsevier, Amsterdam, The Netherlands, pp. 37–66, 2007.

[RUA 02] RUAS A. (ed.), *Généralisation et Représentation multiple*, Traité IGAT, série Géomatique, Hermès-Lavoisier, Paris, 2002.

[SES 99] SESTER M., *Automatische Generalisierung mittels Ausgleichun*, No. 17, Mitteilungen des Bundesamtes für Kartographie und Geodäsie, Verlag des Bundesamtes für Kartographie und Geodäsie (BKG), Berlin, 1999.

Part 2
Summary and Suggestions

Chapter 10

Analysis of the Specificities of Software Development in Geomatics Research

This chapter offers a synthetic analysis of the specificities of the work that has been presented: AROM-ST, an object-oriented knowledge representation language devoted to the representation of spatiotemporal phenomena; GENGHIS, an application to design spatiotemporal information systems in the fields of risk which rely on this language; GENEXP-LANDSITES, an application to generate space tessellations which present specific characteristics; GEOLIS, an application for geographical data exploration-querying; GEOXYGENE, a platform to develop object-orientated processes on geographical data; MDWEB, a module to catalog data and geographical services; ORBISGIS, a GIS focused on the city; and WEBGEN, a generalization Web service platform.

Chapter written by Florence LE BER and Bénédicte BUCHER.

We would like to organize this analysis along the headings of the description canvas of these projects which have proven, after analysis, the most relevant to identify the project classes: origin and motivations, major functionalities, and reusability (architecture and field).

10.1. Origin and motivations

10.1.1. *Targeted users and uses*

The projects presented in Chapters 2 to 9 can be organized into two major categories, depending on whether specific users were targeted at the beginning of the project, and on whether there are developers participating in the project or not.

Thus, GenExP-LandSiTes, Geolis, MDweb, and OrbisGIS aim for precise thematic uses for users who are not direct contributors to the tool itself. More specifically, GenExP-LandSiTes is meant for agronomy and agroecology researchers; Geolis for geographers; MDweb for thematic mapper technicians using and producing geographical data on a single territory; and OrbisGIS for city thematic mapper technicians. GenGHIS is also a part of this category, since it is designed for information system designed in the field of risk, and not for GenGHIS contributors.

Arom-ST, GeOxygene, and WebGen are mainly meant for participants to the development of the tools or for researchers and engineers in the same field. Arom and Arom-ST, for instance, are usable by the community of researchers working on knowledge representation languages. The two projects GeOxygene and WebGen have for their part evolved to also be open to outside users: GeOxygene's matching module is used outside of a matching research context, and WebGen is used by the industrial sector.

10.1.2. *Motivations and foundations*

As for strategic motivations, we have identified two main categories depending on the dominating motivation: projects (1) which aim to drive computer science forward and for which geomatics is rather considered as a field of application which presents new challenges, and projects; (2) which are more thematic and aim to adapt computer science techniques to solve geomatics issues.

In the first category, we have the AROM-ST and GEOLIS projects. AROM-ST's goal is to drive the field of object-oriented representation languages forward, by giving them elements devoted to the expression and the manipulation of spatial characters. The GEOLIS developers want to progress in the field of data exploration. Let us remind the reader that these projects are the projects that show the highest level of declarative representation of knowledge, in compliance with the symbolic artificial intelligence principles at work here. That said, in the specific case of spatial analysis, algorithmic geometry is a field in which declarative representation of knowledge is still hard to apply.

The second category is made up of projects which aim less to drive computer science forward and more to solve a geomatics issue. Let us remind the reader that these projects often opt for an innovative computer science technology: the Web services for WEBGEN, the Web services and cataloguing metadata for MDWEB, the object orientation and the agent-based modeling for GEOXYGENE, and the AROM-ST language for GENGHIS. Among these projects, GEOXYGENE aims for a computer science modeling with a certain level of expressiveness and MDWEB also integrates ontologies. In this latter category, we can distinguish a certain number of projects which we will qualify for development integration, since they aim first to build on developments within a community and then to share them. GEOXYGENE, thus, aims first to integrate and build on proposals designed in the COGIT laboratory. Its mission to transfer the knowledge outside and extend the community with scientific collaboration came after, as a natural consequence. WEBGEN aims to integrate generalization tools, designed within a clearly identified researcher community, and make them interoperable and comparable. As for ORBISGIS, it was designed to build and capitalize on the proposals for urban information systems carried out at the IRSTV.

Let us also remind the reader that more and more developments are carried out or changed under the double impetus of queries from thematic mapper technicians and breakthroughs in terms of model design and computer science development. This is particularly the case for GENGHIS that aims both to promote AROM-ST and to deal with an important geomatics issue. This is also the case for WEBGEN and MDWEB. The GENEXP-LANDSITES project also aims to answer a specific thematic question and leads to joint progress in thematic, statistic, and computer science research.

10.2. Major functionalities, fields, and reusability

10.2.1. *Functionalities*

As for functionalities, the projects presented are generally found in one of the two following categories. We have not classed AROM-ST which is a language and whose main functionality is thus to help implement software.

The "GIS platform" type projects aim to integrate most of the functionalities of a GIS into various modules (which can be imported), by developing them more or less depending on the platform (ORBISGIS and GEOXYGENE).

Other projects focus on a classical functionality of GIS but decline it innovatively. GEOLIS studies the classical geographical data exploration and query functions, but offers to implement them in an innovative way through a logical expression system. GENGHIS focuses on the spatiotemporal data visualization and query functions but implements them innovatively in many ways (Web, etc.). WEBGEN offers analysis functions (generalization) but allows the user (of the GIS client) to use them remotely. GEOXYGENE aims to offer functions which integrated the IGN expertise in vector data management into automatisms.

Finally, the other projects offer functions which currently (in 2012) do not exist in the definition of a GIS and will probably become core functionalities in future GIS. The GENEXP-LANDSITES project thus offers to generate data with the desired characteristics, allowing us to carry out sensitivity studies (test data) and long-term planning. The MDWEB project offers to research geographical data or geographical services online.

10.2.2. *Fields*

The notion of "field" in geomatics can refer to application fields (hydrography, sustainable city, etc.) or methodological fields (artificial intelligence, remote detection, etc.). A single field is sometimes considered like an application field – for instance, cartography is an application field of generalization – and a technical field – for instance, cartography is a technical field for many GIS projects.

Some projects are clearly positioned on thematic fields and others more on methodological fields. Thus, ORBISGIS is essentially positioned on the city field, and makes, in a lesser way, the methodological choice of raster modeling. GEOLIS is clearly positioned on a methodological field, the field of logical information systems. WEBGEN and MDWEB are positioned strongly in technical (generalization, knowledge sharing) and methodological (Web services and Web service cataloguing) fields. GENEXP-LANDSITES is also strongly thematic and technical (generating field patterns through spatial tessellation methods which are in the field of algorithmic geometry and spatial statistics). GEOXYGENE is mainly positioned on the methodological side (object-oriented programming and geographical data vector modeling).

A project's field is important for pooling. Two projects covering the same thematic field can look for function complementarity (build a new process using the functions of one and the other), method comparison or network pooling. Two projects with the same technical field can look for complementarity in terms of method comparison and application comparison: improving our knowledge of a technique by learning how it is applied to other issues, assessing the applicability of our own methods to other thematic fields. For instance, there is a specific literature focusing on the application of agent-based techniques in the spatial field where researchers can confront their proposals.

The restrictions that a field brings to a piece of software's reusability seem to be threefold:

– The software interface (programming or graphical) might require expertise inherent to the technical field to be used. For example, the COLORBREWER library in GEOTOOLS, devoted to the offering of ranges of color to write thematic maps, requires the user to understand the notion of sequential, differential, and ordered relation. In the same way, to correctly set the online accessible processes of the WEBGEN platform, we must understand the meaning of the parameters to assess them accurately.

– The data model can be very specific to the field. For example, an itinerary calculation tool that takes in input all the transportation networks and interconnections as well as the user's preferences (on foot, etc.) seems to be extremely specific to that field. However, a tool to calculate the shortest path is still very generic. On the contrary, in GENEXP-LANDSITES, an extremely narrow project, the methods used are spatial statistics and tessellation methods which are generic and can be applied to objects different from agricultural fields.

– The work distribution network. There are cases in which a piece of software using a certain technique to deal with an issue in a certain thematic field will be completely unknown to researchers working in a different field and using different techniques, when it could be of interest. Let us imagine, for instance, that a researcher wishes to transmit hydrographic data over a mobile network. He/she can make relevant use of road network generalization methods, extrapolate them into his/her water system network to simplify it and transmit it in a more simple manner so that it takes up less of the client's memory.

10.2.3. *Reusability*

Reusing a development may differ depending on the project. Beyond the reusability limits linked to the specificity of a field, which was mentioned previously, the reusability depends on the choice of models and architecture, language, open source or proprietary aspect, or even the coding quality.

AROM-ST initially had a very low level of interoperability since it was a new language, but its authors aim for its conversion into OWL, which has become, in the last few years, the new standard in knowledge representation on the Web. As for GENGHIS, it is interoperable within a small measure, on the level of the result, since it uses standard Web technologies and the generated interface can thus be retrieved and modified.

GenExP-LandSiTes is meant to be coupled with software which loads space tessellations to simulate agroecological processes and focuses the issue of interoperability of these client modules and their input format.

Geolis can read GML and be used remotely (independent of the client platform). That said, to use Geolis's capacities to their fullest, we must use its language to describe its field.

GeOxygene and OrbisGIS were both designed as interoperable platforms inasmuch as the data models used are ISO norms. This interoperability is however limited at the model design level since during the development there was no norm implementation. The data coupling can happen due to XML implements but the API coupling is more complex. That said, OrbisGIS does not implement certain complex elements of the ISO models and is thus more interoperable. Both offer extension mechanisms to their interface.

MDweb is the project which can be most reused among all the projects proposed in this book, to us. It has been designed as a component for an open and extensible modular architecture. It focuses on specific functions and deliberately relies on standard model implements, whether in terms of manipulated data formats or of interface, i.e. of proposed functionalities.

WebGen belongs to a context of sharing and reusing processes and Web services. It was built around an *ad hoc* communication model which then evolved to take up OGC proposals in terms of geographical data processing service description.

Chapter 11

Challenges and Proposals for Software Development Pooling in Geomatics

This chapter aims to draw up a list of the various issues specific to resource pooling in geomatics and to suggest a few ideas to solve them. We use the term "pooling" to refer to any grouping of resources that allows us to lower development costs and improve scientific progress. During the design and development activities that we have presented, there are various types of resources which can be pooled: the abstract models and the corresponding expertise, and the implementation models and the corresponding components. This chapter is divided into two parts: the first dealing with the requirements and challenges and the second with the suggestions of solution.

Chapter written by Bénédicte BUCHER, Julien GAFFURI, Florence LE BER and Thérèse LIBOUREL.

11.1. Requirements and challenges

In this section we will describe the main requirements of pooling and analyze the obstacles to it, first in terms of function and component implementation, then in terms of abstract models.

11.1.1. *Pooling function implementations*

A first requirement is to pool function implementations. This requirement creates the classical challenges of interoperability. Some have already been solved because of the progress in modular architecture and norms facilitating interoperability in the field of geographical information, which we will summarize in section 11.1.1.1. Challenges remain and will be detailed in sections 11.1.1.2 and 11.1.1.3.

11.1.1.1. *Reusing functions implemented in geomatics*

Manipulating geographical data requires many functions, as we mentioned in Chapter 1 (e.g. loading data, modeling, integrating different data among themselves, visualizing). During the design phase of a new geomatics software project, some of the functions are not targeted by innovation. We should then look at the pros and cons of an *ad hoc* implementation or reuse.

Reusing implemented functions is generally encouraged by the current breakthroughs in reusable software design. Indeed, the reuse of an implemented function requires us to have an access interface to this function. In this case we can talk of an interface in a very broad sense of communication ability with a piece of software: when it comes to what we can ask and what kind of answer we get, a key element is the interface contract specifying the possible interaction with the piece of software. And in the past years, we have seen a proliferation in the development of software providing interface contracts to interact with them on a programming level: software that can be configured with a simple file, libraries offering programming interfaces, libraries offering extension points, and *open source* libraries. Moreover, normalized exchange formats of UML model representation, such as XMI (XML *Metadata Interchange*), also help integrate existing model components within new models.

More specifically, in geomatics, reusing implemented functions is also encouraged by the breakthroughs in interoperability between software manipulating geographical data, as we have seen in the introduction to this book (section 1.2.2). To be more precise, interoperability refers to the use of conventions to help communication. These conventions concern the content exchanged, the requested functions, and the interaction mode to obtain the function. In the case of the exchanged contents, there are *de facto* standards which have been around for a long time to exchange geographical data between components. The most famous one is ESRI's *shapefile* format. In the past few years, the work of the ISO/OGC have led to defining more evolved norms to structure geographical data. There are metamodels to model data (such as Feature [OGC 09]) and a XML implementation (GML or GeoRSS). As for functions and interaction modes, the work of the ISO/OGC has also focused on providing precise functions (data query, catalog query, and data drawing) and more precisely providing it through a Web service type interface. The most famous interface specifications are the Web Map Service (WMS), Web Feature Service (WFS), Web Coverage Service (WCS) and Catalog Services for the Web (CSW) norms. Not only does the norm help reuse existing components but it also discourages *ad hoc* implementation. Finally, geomatics norms are often complex and the cost of the implementation incites people to reuse pre-existing interoperable components rather than reimplement these norms.

Some pieces of software already correspond in part to this need for the most elementary functions in a GIS project. Most of the developments which require storing in a database management system (DBMS) rely on POSTGRESQL or ORACLE. The JAVA pieces of software manipulating geographical data often reuse the JAVA Topology Suite to represent geometries and carry out a few elementary operations on these geometries (intersections, convex hull, etc.). In the Web domain, due to the dissemination of ISO/OGC norms previously mentioned, some implementations have been proposed, such as the MAPSERVER, GEOSERVER, GEONETWORK, or even MDWEB pieces of software, which are widely used. Among the projects presented in this book, ORBISGIS uses MDWEB components for the cataloging part, to benefit from these functionalities while sparing the cost of implementing relatively complex norms (CSW norms). Moreover, the initiatives to facilitate free access to a base map and data location on this base map have been successful. For example, we can mention Google API or Geoportail's API. In the field of non-Web application development, some GIS, such as UDIG, QGIS, or OpenJUMP, offer extension points to reuse their graphical interface.

The display and interaction functions *via* a graphical interface (select, edit, etc.) are also often reused since their development requires specific knowledge (2D JAVA libraries, for instance), which is often different from the analysis expertise that certain teams have, and their development is relatively costly. In this case, there is a triple advantage. The reuser has access both to the drawing functions as well as to the function provided as *plugins* with this tool. He/she can also choose to distribute his/her function to other users of the tool by providing them with his/her *plugin*. So OpenJUMP users can easily connect to the WEBGEN services because of a dedicated *plugin* and the WEBGEN community can connect to other works published in OpenJUMP.

11.1.1.2. *The challenge of defining interoperable interfaces*

The breakthroughs in interoperability should not obfuscate the issues which remain when it comes to exchanging content and functions.

In the case of content, as we have explained it in the introduction of this book (section 1.1.1), a specificity of geographical data is the lack of consensual model to structure it. The work of organizations like ISO and OGC is a considerable improvement for interoperability but still has its limits.

On the one hand, the current norms do not cover enough modeling elements. Many mathematical elements involved in models (point, line, and polygons) are covered by abstract norms. Yet there are many complex structures used in geomatics, such as trees, graphs, networks, and partitions. Moreover, there is a lack of consensus about other non-mathematical elements, such as models of lakes, roads, or buildings, and, to exchange these elements, the norms must allow us to share not only the data but also the model definitions. And the current norms for geographical metadata do not currently allow us to formally transcribe a model's entire semantics.

Additionally, the interoperability of the content exchanged in random access memory is not guaranteed by the norms which limit themselves to XML implementation (serialization). Thus the different JAVA implementations of Feature (such as the implementation in GEOTOOLS, the implementation in DEEGREE, and the implementation in GEOXYGENE) are not interoperable. However, this is an area that requires a very large amount of work: a great number of adaptors must be encoded because many classes are embedded one in the other to describe complex geometry. This is the point of works being carried out to define a GEOAPI, a set of JAVA interface functions implementing ISO/OGC models to describe geometries and objects.

In our view, the most controversial issue is the description of functionalities proposed to promote their reuse [LEM 06]. We will mention here two types of function description necessary to pooling.

The first type is the detailed description of input and output of the function, including pre-conditions and post-conditions, the conditions and effects on the manipulated representation. When pooling means that an application automatically parses the description of a function to call it up or to combine it with another function, this application needs this first type of description. It implies on the one hand a clear identification of input and output, and on the other hand, the use of logical formalism to write them formally. A substantial obstacle here is how we can find and share what we can call "the function's validity context". This refers to the requirements concerning the input data for the tool to actually provide the described functionality. These conditions are not always limited to being compliant to the specified input format. The difficulty is double in this case: knowing the conditions and expressing them without ambiguity. The process is generally developed and tested in a limited context (on specific data). There is no unitary testing in geographical information [BUC 07, BUC 08]. Moreover, non-ambiguous communication of these conditions to an application requires models with formal semantics able to express these wordings (KIF[1], SWRL[2], etc.).

The second type of function description is the designation of its global effect. This type is often necessary when the

---

[1] http://www.ksl.stanford.edu/knowledge-sharing/kif/
[2] http://www.w3.org/Submission/SWRL/

pooling involves experts in a field since they must be able to identify a function with an understandable term (drawing, simplification, and calibration) and not with the formal description of its input and output. There is no consensus on typology in this field. [BER 87] and [TOM 91] proposed a data operation classification adapted to *raster* or unique data. Their underpinned goal is to arrive at a general methodology for thematic data analysis. [MIT 99] also proposed a task and subtask model to formalize the process of thematic map writing. [LIN 10] proposed an environment to help scientists design and validate processing chains based on the expert knowledge of manipulated data and process categories. [GIO 94] and [ALB 96] proposed a GIS operation classification aiming to facilitate the design of processes above various GIS tools: a process is described through abstract operations and can be carried out in different GIS tools offering different implementations of the operations described. There is also a very high-level ISO taxonomy (ISO 19119) that has, for example, a *geoprocessing services* category subdivided into four categories according to the nature of the information processed: spatial, thematic, temporal, and metadata. The OGC consortium that wishes to define interface specifications for geographical data processing services called to the definitions of specialized profiles for Web processing services [OGC 07a, OGC 07b]

11.1.1.3. *The challenge of modular development*

Another obstacle to function implementation pooling is that the techniques of modular programming are not yet widely used. Few researchers design functions as modules reusable by others, meaning that few of them grasp the fact that their functions could be useful to others and make the effort of implementing functions as reusable components. Yet, when a function is not designed in a modular fashion, trying to extract it retroactively from its original application by copying the method query sequence which matches it requires us to know the variable used by these methods to document them accurately. This experiment was carried out in the COGIT laboratory[3] to design a service that automatically improves color contrast from a method embedded in an existing applications. The authors analyzed the difficulties they came across to extract a method from its original context so as to provide it as a service [BUC 08]. We can explain the lack of modular development by the fact that these approaches are still new and sometimes complicated to establish. Moreover, this investment has little scientific value in the short term.

In addition, an increasing issue that goes against reusing code in a JAVA development context is the evolution of widespread libraries, such as the XERCES library for XML data management in JAVA or the GEOTOOLS library, of which different versions are not always compatible. When such libraries are frequently used, it can create conflicts when reusing a component in a development. The two components cannot be executed in the same process since they use incompatible versions of the same libraries. This issue is solved with interprocess communication and more generally with the Web-based programming which allows us to avoid platform incompatibilities.

Finally, when it comes to the source code, the issue of the exercise is to first understand a code, a program structure, and the detail of the methods. The current literature is still insufficient. A JAVA program literature is typically organized according to the *packages* and classes. However, this organization is not always the organization of the functions. What is more, using code commentaries can create a language problem. This caused a delay in the adoption of the *open*

---

3 http://recherche.ign.fr/labos/cogit/accueilCOGIT.php

source GVSIG piece of software by non-Spanish speakers. A last challenge in code pooling is to help find more input from developers taking up an *open source* code. Globally, we should accept lifecycles that alternate divergence and integration phases. During the divergence phases, subprojects appear and must be self-sufficient enough to explore an aspect of software development without being constrained by other aspects. During the integration phases, the divergence must be reduced to the minimum, when possible, to allow for efficient integration of various building blocks. The concordance of both phases is delicate but divergence is necessary for code progression. In a more general way, this is the issue of collaborative publishing of models, based for instance on collaborative development tools such as SVN [MIC 11].

11.1.2. *Pooling models and expertise*

11.1.2.1. *The need for it*

Another type of resource that can be shared covers the models and algorithms used. For example, the 3D building generalization algorithm proposed by Martin Kada [KAD 02] was taken up and improved in GEOXYGENE by using the description he made of it in a scientific article. GENEXP-LANDSITES codes tessellation algorithms described in literature and normally developed for other uses [ADA 07]. AROM-ST is derived from the AROM language, itself developed on the basis of many different French works on knowledge representation [DUC 98]. GEOLIS, specific to geolocated data, is built from a generic data storing and querying system [FER 04]. To share these resources, it is important to define a set of framework articles which described the models and which can be quoted. Moreover, the access to a demonstrator or prototype is an advantage that helps pooling. There are many structures which help this kind of pooling since it is at the heart of research's progress: conferences, workshops, journals, and reference sites (CITEULIKE, ARXIV, HAL, and ORI-OAI). Let us mention that in the specific field of geographical information, this is the most important type of pooling. Indeed, this field is multidisciplinary and its breakthroughs are often due to teams whose members have different fields of competences learning to share and integrate their models and methods.

Another type of resource is the expertise used during the development: user implication method, library ownerships, and design patterns. Pooling these resources means publishing and reading scientific articles, possibly in the fields which are not connected to geomatics (such as ergonomics, work psychology, sociology, or cognitive anthropology), exchanges during conference and workshops or on forums. It can also rely on the transfer of people, such as a postdoctoral candidate or a contributor. The establishment of collaborative projects, under the impulse of funding from the European Commission or the French National Research Agency, among others, is an efficient way to pool methods and development in promoting team collaboration.

11.1.2.2. *A challenge: the diversity and gaps in the existing expertise*

The final obstacle to resource pooling in geomatics seems to be the difficulty we have in knowing what is existing – what has been done, what can be reused, and how. This difficulty is the consequence, on the one hand, of the abundant and heterogeneous aspect of the existing (a lot of things but they are hard to grasp and compare), and, on the other hand, of the diversity of necessary expertises. As we underlined it in the introduction of this book, geomatics is a scientific field at the crossroads of other fields (such as cartography, databases, mathematics, and geography). Many competences make up geomatics. It appears that geomatics has built itself in the past few years on various fields whose vocabularies were always separate. Its vocabulary is thus particularly ambiguous since it refers to various original disciplines without explicitly mentioning them. When we talk of a building model, for example, is it a conceptual diagram or a set of data representing a building or even a simulation method reconstructing a building from photographs? This ambiguity is hard to avoid even for similar disciplines and this book has encountered a great amount of it. Moreover, geomatics is a rapidly changing field along with information and communication technologies, and it must integrate emerging themes such as information infrastructures, localized services, or even geographical information retrieval. Its vocabulary is permanently being rebuilt to adapt to technological evolutions and research breakthroughs. Finally, this field still has not been identified by scientists who do not belong to it but might encounter issues linked to this field. All of this makes pooling difficult.

Finally, in the case when we pool software resources without pooling expertise, we might end up with processing or geographical data misuses. This use does indeed often require a certain level of expertise both in the application of the tool and the analysis of the results. The data or function provider might fear that bad decisions might be made due to the resources he/she provided because he/she will not have been able to accompany the process built on his/her resources.

11.2. Solutions

The previous section showed the main challenges for pooling resources useful to geomatics software development. This section will confront us with interoperability, tool and algorithms cataloging, and model indexation issues in scientific communities and to the issues of communication from one community to the other.

This section presents existing solutions or solutions currently being developed to remove some of these impediments.

11.2.1. *Reference frameworks and metadata*

A key concept for pooling is the concept of a reference framework. From what we have said previously, we can deduce that three types of reference frameworks are needed:

– design frameworks allow us to build the most interoperable components possible;

– description and cataloging frameworks;

– competence frameworks for the visibility of scientific communities facilitating knowledge pooling.

The first solutions go the way of providing frameworks for resource design, typically specification implemented with characteristic interfaces in a given domain which the designers are intended to follow so that their code will be interoperable. GEOAPI and the WEBGEN community work on the precise definition of structures that can be shared, want to facilitate the sharing between libraries and the design of interoperable services. We will classify the MOBIDIC platform among these solutions, a platform found in the context of mobile service development [LOP 09]. The authors allow the developers of localized Web services or other software components to integrate into their services context elements (not only the user's location but also the size of the roaming device's scree and memory). The values of these context elements are acquired due to the MOBIDIC middleware present on the client side.

There is other work currently focused on building semantic frameworks to describe components to ease their cataloging and use: thesauri, taxonomies, or ontologies. These frameworks are useful inasmuch as they are used to build resources (for an attribute value domain, for example), and especially when building metadata. The projects which aim to facilitate the construction and publication of metadata are a key element in pooling.

In general, Web services are very fond of frameworks to describe their interface and do their cataloging. Indeed, there are already existing frameworks for their description, such as the standard Web Service Description Language (WSDL) and frameworks for their cataloging (such as ebXML and ebRIM). In geomatics, Lutz and Klien [LUT 06] used an ontology to reuse specific Web services which are geographical object providing services (in the WFS standard). Thus their ontology is centered on the description of geographical objects. In the same way, WEBGEN is centered on the design of reusable generalization Web services and also deals with service description. Burghardt *et al.* [BUR 05] propose precise typologies for generalization processing services. Their work underlines the need to use various classifications to describe a functionality and list processing services. Thus one of the proposed classifications organizes processes according to their signatures, i.e. according to the input and output data types; another classification organizes the processes according to the functional components of a generalization process (data preparation, generalization operations, and process control). [LEM 06] proposes various ontologies to catalog and facilitate the design of processing services: an ontology of real-world objects, an ontology of database objects, an ontology of atomic operations, and an ontology of complex operations. His proposal insists on allowing us to check the consistency of a processing chain at the input and output data level. [TOU 10] proposes an ontology to describe application conditions and generalization process effects to organize different generalization processes in a heterogeneous space. The MOBIDIC platform previously mentioned also proposes a service description language that promotes the discovery and assembly of mobile services developed with this platform [LOP 09]. This model, called WSRM, enriches the Web service description standards with the service application domain (such as online learning) and the service use context (which can be a client- and user-side prerequisite in terms of user location and client physical characteristics).

That said, reusable resources are not all Web services. For example, in [BUC 08], the authors focus on the issue of reusing processes embedded in applications which have not been designed as Web services. The authors take up a principle of the *Service Oriented Architecture* approach: a relevant concept on which software reusability can be grounded is the concept of the services provided by the software. From there, they suggest a specific metadata model, called model and function metadata (MFM), to describe services provided by software resources and how to access these services. Their proposal does not impose a framework for the metadata literature (e.g. the function taxonomy) but offers the designers the possibility of building and exchanging the frameworks which they will use in their descriptions. In the field of process descriptions, a pioneering work was written by [BEC 02] who proposed a model for urban model interconnections. In this case, the expression *urban model* refers to a process allowing us to calculate, with known quantities, quantities that we are looking to simulate or approach. [BEC 02]'s proposal is to match each module with an adaptor complying to a common description model. His proposal also helps carry out a complex process that uses various urban models in a synchronized manner, which can be transposed to engineering in the context of Web services. Since then, design methods consisting of specifying components through interface contracts before implementing them have also helped reuse their component functions and their composition.

The work on the semantic Web or on the manipulation of spatiotemporal data also goes in the direction of helping to share documents or data available on the Web by matching them with structured metadata complying to non-ambiguous models. They focus on data and do not deal at all with service integration.

Finally, when it comes to improving the visibility of domains of expertise, and in the long term, scientific communities, let us mention the work in progress of American researchers in geomatics: a group of American researchers, each a key researcher in their own field, has established a corpus of knowledge in geographical computer science which have around a hundred sections, called the *GIS&T Body of Knowledge* (GIS&T BoK) [DIB 07]. Actors of the GIS&T BoK are spreading out their work to the international community and are contacting the main associations in the field of geographical computer sciences to validate this framework and eventually adapt it to specific context. Other initiatives are multiplying to provide greater visibility to geomatics training, whether this is for future students or for employers. A survey was carried out in France, supported by the MAGIS research group and the AFIGEO association, to study geomatics in French. In Europe, the authors of [RIP 11] have also studied and compared various ways to teach geomatics.

11.2.2. *Test cases to improve description of implemented functions and progress within a community*

A promising proposal for resource pooling in a research community, which was debated within the Pooling project of the MAGIS research group and within the WEBGEN community, is the definition of a more or less large set of test cases. For example, the 3D computer graphics researchers working on illuminance models use a common 3D set often called "Cornell box" to compare their illuminance models (see Figure 11.1). This set is made up of a box in which there are relatively simple objects. The data for this test case are available to everyone [COR 98].

Figure 11.1. *The Cornell box*

Sharing these common test cases holds various advantages for a research community:

– It facilitates the comparison of various approaches (such as, for instance, the different types of illuminances models: the display of their effect on the Cornell box allows us to get a sufficient idea of their differences).

– It facilitates the data access for the approach test. This is especially true in fields where the question of data acquisition is an issue.

– It promotes communication within the research community and outside it, allowing researchers to communicate about their work on use cases that might be more significant for the audience (or at least cover a larger spectrum). In particular, this makes it much easier to teach the goals and techniques of a specific field of research to junior researchers or students.

– It promotes collaboration between researchers and users: the users can suggest test cases in which they take interest.

– It enables the identification of a large number of issues which the research community more or less deliberately allows to slide.

– It promotes the specification of some functions. These functions sometime require concrete examples to be clearly communicated (such as functional tests). This is in particular the case for cartographic functions.

Various initiatives have been carried out on the sharing of test cases. For example, in information searching, test campaigns were organized for search engines focusing on identical issues. The reason for organizing these campaigns was the cost linked to the construction of test sets (sets of queries and relevance assessments) in this field. The test campaigns allowed us to compare, in the most unbiased way possible, these engines and helped the research progress in this field [GEY 06, PAL 10].

The research community in automatic generalization also developed a project called European Spatial Data Research Network (EuroSDR) focusing on the assessment of automatic generalization software [BUR 08, STÖ 08, STÖ 09]. The software was assessed on identical generalization test cases proposed by different European cartography agencies; some of them are mentioned in Figure 11.2. The first test case was

310 Innovative Software Development in GIS

provided by the Catalonia Cartography Institute, the second by the United Kingdom's Ordnance Survey, the third by the Netherlands' Cartographic Institute, and the fourth by the French National Geography Institute.

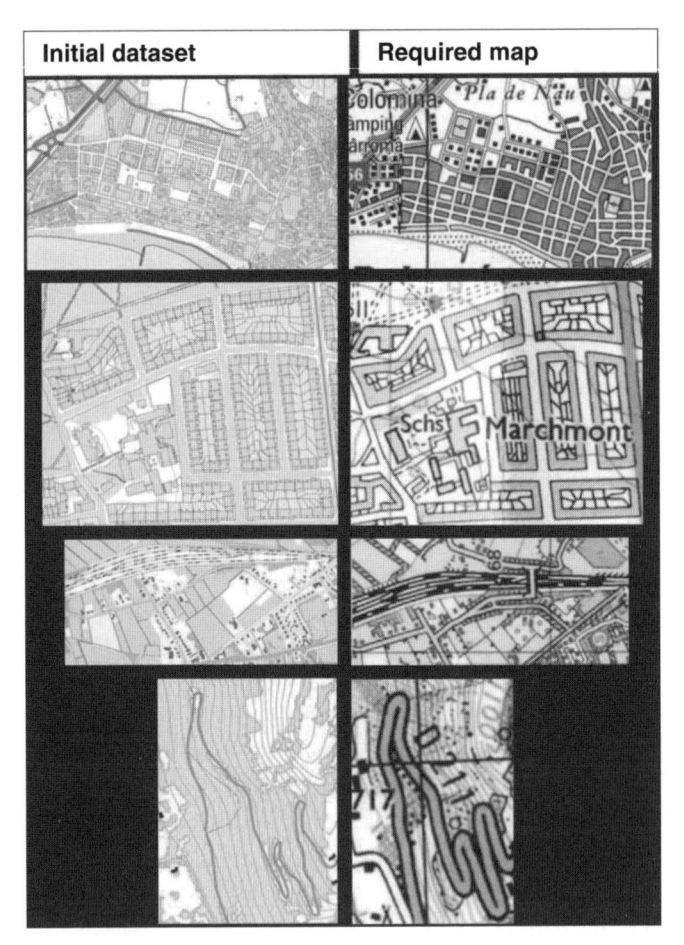

Figure 11.2. *Generalization test cases examples*

To design a satisfying set of test cases, we must respect as best as possible two opposite constraints. On the one hand, we must have a sufficiently low number of examples, to ensure that each of them is known and used by a maximum of persons. On the other hand, we must have a sufficiently large number of examples, to ensure that all the types of issues are covered. For example, if more generalization test cases could be proposed, they would have to cover as best as possible the criteria space allowing us to qualify a generalization issue type. These criteria can be:

– the types of data considered (buildings, buildings and roads, weather data, etc.);

– the magnitude of the detail level change;

– the type of geographical landscape in question (urban, rural, coastal, mountainous area, etc.);

– the domain (hiking map, road map, planning map, etc.);

– the size of the data (2D, 3D, spatiotemporal data, etc.).

This list is obviously not exhaustive and the criteria mentioned are not independent, one from another. If all the tests cover enough of the possible issues, a user with a given issue will be able to find a test case that will be similar to them, and thus have an idea of the research results that might interest them.

11.3. Conclusion

The need for pooling is strong in geomatics. It goes beyond the field of software developments studied in this book. The main pivots are the sharing of frameworks: frameworks to design resources (essentially API), frameworks for description (ontologies), and frameworks for the positioning of a scientific field. It is also essential to keep a critical eye on the norms, since they have a key role in pooling and their establishment and dissemination cannot be solely based on the needs of different lobbies. Moreover, frameworks are not enough. Two types of tools must be linked to the frameworks to facilitate pooling: tools helping to build and validate metadata and tools helping to use metadata to search for resources and use them. The first type should still be the focus of proposals and we believe that the designing of test cases and test games is very promising in this respect. The second type of tool is the catalog, represented in this book by the MDWEB project (Chapter 6).

Beyond function, model, or component pooling, expertise pooling is a source of innovation. To this end, the pooling must integrate different points of view: that of the researchers, the thematicians (geographers, archaeologies, geomorphologists, etc.), or the computer scientists (database or image processing specialists, as well as knowledge engineering specialists); that of the users (land planners, land managers [LAR 05], even drivers or hikers [DOM 10]); and that of geomaticians, the latter having a specific role as broker between different disciplines. Thus the GENGHIS project presented in the book (Chapter 5) brought together computer scientists, geomaticians, and mountain risk specialists to develop a tool based on a spatial and temporal knowledge representation system, the AROM-ST system. In the same way, the GEOLIS project (Chapter 6) gathered computer scientists and geographers to offer a geographical data querying system based on a logical information system.

There are other techniques to explore jointly and they can promote innovation in geomatics: let us mention, without claiming to be exhaustive, qualitative reasoning models over time and space developed in artificial intelligence [LEB 07, LIG 10], analogy reasoning models (such as case-based reasoning), current models of knowledge representation (conceptual graphs, description logics), and ontology development techniques, or even the different approaches developed within the framework of information systems. Innovation is also promoted when we are better at taking into account the geomatics users and uses, as [NOU 09] suggested recently. This is the way geomatics stand to gain from a better integration of social and cognitive science research work.

11.4. Bibliography

[ADA 07] ADAMCZYK K., ANGEVIN F., COLBACH N., LAVIGNE C., LE BER F., MARI J.-F., "GenExP, un logiciel simulateur de paysages agricoles pour l'étude de la diffusion de transgènes", *Revue Internationale de Géomatique*, vol. 17, nos. 3–4, pp. 469–487, 2007.

[ALB 96] ALBRECHT J., "Universal GIS operations for environmental modeling", *Proceedings of the 3rd International Conference on Integrating GIS and Environmental Modeling*, Santa Barbara, USA, 1996.

[BEC 02] BECAM A., YEHUDI: Un environnement pour l'interopérabilité de modèles urbains distribués et homogènes, PhD in Computer Science and Information Society, INSA Lyon, 2002.

[BER 87] BERRY J., "Fundamental operations in computer-assisted map analysis", *IJGIS*, vol. 1, no. 2, pp. 119–136, 1987.

[BIS 97] BISHR Y., Semantic aspects of interoperable GIS, PhD Thesis, ITC, The Netherlands, 1997.

[BUC 07] BUCHER B., BALLEY S., "A generic preprocessing service for more usable data processing services", *Proceedings of the 10th AGILE Conference*, Aalborg, Denmark, 2007.

[BUC 08] BUCHER B., JOLIVET L., "Acquiring service oriented descriptions of GI processing software from experts", *Proceedings of the 11th AGILE Conference*, Girona, Spain, 2008.

[BUR 05] BURGHARDT D., NEUN M., WEIBEL R., "Generalization services on the web – a classification and an initial prototype implementation", *CaGIS*, 2005.

[BUR 08] BURGHARDT D., SCHMID S., DUCHÊNE C., STÖTER J., BAELLA B., REGNAULD N., TOUYA G., "Methodologies for the evaluation of generalised data derived with commercial available generalisation systems", *11th ICA Workshop on Generalisation and Multiple Representation*, Montpellier, France, 2008.

[COR 98] Cornell University Program of Computer Graphics, The Cornell Box (online), 1998, available at http://www.graphics.cornell.edu/online/box/.

[DIB 07] DIBIASE D., DEMERS M., JOHNSON A., KEMP K., LUCK A., PLEWE B., WENTZ E., "Introducing the first edition of the GIS&T Body of Knowledge", *Cartography and Geographic Information Science*, vol. 34, no. 2, pp. 113–120, 2007.

[DOM 10] DOMINGUÈS C., BALDIT-SCHNELLER P., "Les randonneurs définissent leurs cartes. Exploitation d'une enquête semi-directive à questions ouvertes avec des outils statistiques et linguistiques", *SAGEO'10, Spatial Analysis and Geomatics*, Toulouse, France, 2010.

[DUC 98] DUCOURNAU R., EUZENAT J., MASINI G., NAPOLI A., (eds), *Langages et modèles à objets – Etat des recherches et perspectives*, Collection Didactique D-019, INRIA, Le Chesnay, France, 1998.

[FER 04] FERRÉ S., RIDOUX O., "Introduction to logical information systems", *Information Processing & Management*, vol. 40, no. 3, pp. 383–419, 2004.

[GEY 06] GEY F., LARSON R., SANDERSON M., JOHO H., CLOUGH P., PETRAS V., "GeoCLEF: the CLEF 2005 cross-language geographic information retrieval track", *CLEF 2005 Proceedings*, LNCS 4022, pp. 908–919, 2006.

[GIO 94] GIORDANO A., VEREGIN H., BORAK E., LANTER D., "A conceptual model of GIS-based spatial analysis", *Cartographica*, vol. 31, no. 4, pp. 44–51, 1994.

[KAD 02] KADA M., "Automatic generalization of 3D building models", *GIS – Geo-Information-Systems, Journal for Spatial Information and Decison Making*, vol. 9, pp. 30–36, 2002.

[LAR 05] LARDON S., PIVETEAU V., "Méthodologie de diagnostic pour le projet de territoire: une approche par les modèles spatiaux", *Géocarrefour*, vol. 80, no. 2, 2005.

[LEB 07] LE BER F., LIGOZAT G., PAPINI O. (eds), *Raisonnements sur l'espace et le temps: des modèles aux applications*, Traité IGAT – Géomatique, Hermes-Lavoisier, Paris, 2007.

[LEM 06] LEMMENS R., Semantic interoperability in distributed geo-service, PhD thesis, ITC, Enschede, The Netherlands, 2006.

[LIG 10] LIGOZAT G., *Raisonnement qualitatif sur le temps et l'espace*, Collection Ingénierie des langues, Hermes-Lavoisier, Paris, 2010.

[LIN 10] LIN Y., PIERKOT C., MOUGENOT I., DESCONNETS J.-C., LIBOUREL T., "A framework to assist environmental information processing", *ICEIS*, pp. 76–89, 2010.

[LOP 09] LOPEZ-VELASCO C., GENSEL J., VILLANOVA-OLIVER M., MARTIN H., "Vers une plate-forme de génération de SIG mobiles adaptés au contexte d'utilisation", *Revue Internationale de Géomatique*, vol. SIG mobiles, 2009.

[LUT 06] LUTZ M., KLIEN E., "Ontology-based retrieval of geographic information", *International Journal of Geographical Information Science*, vol. 20, no. 3, pp. 233–260, 2006.

[MIC 11] MICHAUX J., BLANC X., SUTRA P., SHAPIRO M., "A semantically rich approach for collaborative model edition", *Proceedings of the 26th Symposium on Applied Computing, SAC2011,* Taichung, Taiwan, 2011.

[MIT 99] MITCHELL A., *The ESRI Guide to GIS Analysis, Volume 1: Geographic Patterns & Relationships*, ESRI Press, Redlands, CA, 1999.

[NOU 09] NOUCHER M., La donnée géographique aux frontières des organisations: approche socio-cognitive et systémique de son appropriation, PhD Thesis, EPFL, 2009.

[OGC 07a] OPEN GEOSPATIAL CONSORTIUM, OGC Web Services Initiative – Phase 5 (OWS-5) – Annex B OWS-5 Architecture, 2007.

[OGC 07b] OPEN GEOSPATIAL CONSORTIUM, OpenGIS Web Processing Service – OpenGIS standard, 2007.

[OGC 09] OPEN GEOSPATIAL CONSORTIUM, The OpenGIS Abstract Specification – Topic 5: Features, version 5, 2009.

[PAL 10] PALACIO D., CABANAC G., SALLABERRY C., HUBERT G., "Measuring geographic ir systems effectiveness in digital libraries: evaluation framework and case study", *ECDL'10, 14th European Conference on Research and Advanced Technology for Digital Libraries*, Glasgow, UK, pp. 340–351, 2010.

[PIE 08] PIERREL J.-M., "De la nécessité et de l'intérêt d'une mutualisation informatique des connaissances sur le lexique de notre langue", *Contribution à la table ronde "Lexique" 1er Congrès Mondial de la Linguistique Française*, 9–12 July, Paris, France, pp. 15–30, 2008, available at: http://www.linguistiquefrancaise.org/http://dx.doi.org/10.1051/cmlf08330.

[RIP 11] RIP F., GRINIAS E., KOTZINOS D., "Analysis of quantitative profiles of GI education: towards an analytical basis for EduMapping", *Proceedings of the 14th International Conference on Geographic Information Science (AGILE'11)*, Utrecht, The Netherlands, 2011.

[STÖ 08] STÖTER J., DUCHÊNE C., TOUYA G., BAELLA B., PLA M., ROSENSTAND P., REGNAULD N., UITERMARK H., BURGHARDT D., SCHMID S., ANDERS K.H., DÁVILA F., "A study on the state-of-the-art in automated map generalisation", *11th ICA Workshop on Generalisation and Multiple Representation*, Montpellier, France, 2008.

[STÖ 09] STÖTER J., BURGHARDT D., DUCHÊNE C., BAELLA B., BAKKER N., BLOK C., PLA M., REGNAULD N., TOUYA G., SCHMID S., "Methodology for evaluating automated map generalization in commercial software", *Computers, Environment and Urban Systems*, vol. 33, no. 5, pp. 311–324, 2009.

[TOM 91] TOMLIN C.D., "Cartographic modelling", in MAGUIRE D.J., GOODCHILD M.F., RHIND D. (eds), *Geographical Information Systems: Principles and Application*, vol. 1, Longman, Harlow, pp. 361–374, 1991.

[TOU 10] TOUYA G., "Relevant space partitioning for collaborative generalisation", *12th ICA Workshop on Generalisation and Multiple Representation*, Zürich, Switzerland, 12–13 September 2010.

Glossary

Institutes, organizations, and societies

CEMAGREF, http://www.cemagref.fr/
French Environment Science and Technology Research Institute, recently became IRSTEA "Institut de recherche en sciences et technologie pour l'environnement et l'agriculture".

CIRAD, http://www.cirad.fr/
French International Cooperation Center on Agricultural Research for Development.

CNRS, http://www.cnrs.fr/
French National Scientific Research Center.

IGN, http://www.ign.fr/
Formerly "Institut Géographique National" (French National Mapping Agency), the IGN merged with the National Forest Inventory (IFN) on January 1, 2012 to become the National Institute for Geographical and Forest Information.

INA, http://www.ina.fr/
French National Audiovisual Institute.

INRA, http://www.inra.fr/
French National Agronomic Research Institute.

INRIA, http://www.inria.fr/
French National Institute for Computer Science and Automation Research.

IRD, http://www.ird.fr/
French Institute of Research for Development.

ISO, http://www.iso.org/iso/fr/
ISO, International Organization for Standardization, is an international organization aiming to define norms in various fields.

MAPINFO®, http://www.pbinsight.com/welcome/mapinfo/
MAPINFO, now called Pitney Bowes Business Insight, is a private software publishing company, which has notably published the MAPINFO piece of GIS software.

OGC®, http://www.opengeospatial.org/
OGC, Open Geospatial Consortium, is an international consortium bringing together industrial actors, institutes, and academics aiming to define and promote standards to improve geographical data interoperability and promote software that manipulates these data.

ORACLE, http://www.oracle.com/
ORACLE is a private company specializing in software. One of its products is the ORACLE Database. The ORACLE company bought out SUN and also owns the JAVA language and the database management system MYSQL.

Tools, formats, and languages

API, application programming interface
An API matches a piece of software and corresponds to the procedures of interaction with the program offered to other programs. In this, it is the opposite of the graphical interface (graphical user interface or GUI), exclusively used by humans.

BDTopo®, http://professionnels.ign.fr/ficheProduitCMS.do?idDoc=5287265
BDTopo® is a database from the IGN with topographical objects at a metrical level of detail. Its content is the equivalent of that of a large-scale topographical map (1:25 000 e).

CityGML, http://www.citygml.org/
CityGML is an OGC standard to represent, in XML, data representing a town in 3D.

Deegree, http://www.deegree.org/
Deegree is an open source piece of Java software that provides geographical data and maps on the Web. It also has cataloging functions, which are publishing and querying a metadata base. Deegree complies with the ISO/OGC standards.

DBMS, database management system.

JTS, Java Topology Suite, http://www.vividsolutions.com/jts/JTSHome.htm
JTS is an open source Java library allowing us to manipulate elementary geometrical structures in a 2D Euclidean space (points, lines, and polygons) and to carry out geometric queries on it. These structures follow a standard ISO and OGC model: simple Feature Access Specification for SQL.

GML, Geographic Markup Language
GML is an ISO standard to write geographical data in the XML language.

GeoApi, http://www.opengeospatial.org/standards/geoapi/
GeoApi is an open source Java library offering to implement ISO/OGC models as Java interface to federate Java implementations of these models and make them interoperable. It is, itself, an ISO/OGC standard. The

JAVA interfaces of the GEOAPI are implemented in the GEOTOOLKIT classes and in some GEOTOOLS and GEOXYGENE classes.

GEONETWORK, http://geonetwork-opensource.org/
GEONETWORK is an open source piece of JAVA software that allows us to establish a metadata server that can be queried on the Web. GEONETWORK implements the ISO/OGC standards for these cataloging functions.

GEOSERVER, http://geoserver.org/display/GEOS/
GEOSERVER is an open source piece of JAVA software that uses the GEOTOOLS libraries, allowing it to provide geographical data and maps (overlapping layers obtained by drawing data or stored as images on the server). GEOSERVER implements the main standards of ISO/OGC to provide geographical data and maps online.

GEOTOOLKIT, http://www.geotoolkit.org/
GEOTOOLKIT is an open source JAVA library that implements abstract models of ISO/OGC data and offers different functions such as visualization, server query, and spatial analysis.

GEOTOOLS, http://geotools.org/
GEOTOOLS is an open source JAVA library that implements the abstract models of ISO/OGC data and offers different functions such as visualization, server query, and spatial analysis.

GeoRSS, http://georss.org/Main_Page
RSS is an XML format to exchange synthetic information about a flow of information on the Web. GeoRSS is an XML format that indicates geographical coordinates in an RSS description.

GIS, geographical information system.

GRASS, http://grass.fbk.eu/
GRASS is one of the first open source GIS piece of software, written in the C language.

GVSIG, http://www.gvsig.org/web/
GVSIG is an open source piece of GIS software complying with the main ISO and OGC standards.

IS, information system.

KML, Keyhole Markup Language
KML is a format extending XML to represent geographical data, which is used in widespread tools such as Google Earth and has become an OGC standard (KML version 2.2).

MAPSERVER, http://mapserver.org/
MAPSERVER is a piece of open source software written in the C language that allows us to provide geographical data and maps (overlapping layers obtained by drawing data or stored as images on the server) on the Web. MAPSERVER implements the main ISO/OGC standards to provide geographical data and maps online.

MAPINFO, GIS software edited by the MAPINFO® company

MIF/MID, MapInfo interchange format
MIF/MID is a proprietary data storage format established by the GIS software publishing company MAPINFO®.

OLAP, Online analytical processing.

OpenJUMP, http://www.openjump.org/index.html
OpenJUMP is a piece of open source software using JAVA, allowing us to load, draw, and display geographical data and carry out queries on these data.

ORACLE Database, http://www.oracle.com/us/products/database
Software developed by the ORACLE company. It is a relational database management system with a spatial overlay. This overlay allows it to manipulate geometrical data types, to use operators on geometries. Another ORACLE overlay, database, allows us to manage image types.

OWL, http://www.w3.org/TR/owl-overview/
OWL, Ontology Web Language, is a W3C standard to develop knowledge bases. OWL covers three sublanguages with an increasing expressiveness: OWL Lite, OWL-DL, and OWL Full. OWL-DL works with the framework of description logic and combines the maximum of expressiveness with the completeness of inference procedures.

POSTGRESQL/POSTGIS, http://www.postgresql.org/
POSTGRESQL is an open source relational database management system with an overlay devoted to POSTGIS geographical data. POSTGIS can read and store specific data types to represent geometries, and carry out queries containing operators on these geometries. In particular, POSTGIS allows us to design data spatial indexes.

QGIS or QUANTUM GIS, http://www.qgis.org/
QGIS is an open source piece of GIS software in C++. It is compliant with the ISO/OGC standards.

RDF, http://www.w3.org/RDF/
RDF, Resource Description Framework, is an extremely simple data model used as the basis for semantic Web data models (W3C standard). In this model, any piece of data takes the form of a triolet (resource, appointed by its identifier; property, appointed by a label or an identifier; and literal value or resource, appointed by its identifier).

shapefile (SHP)
The shapefile format is a proprietary geographical data storage format designed by the GIS software publishing company ESRI that has become the *de facto* standard in the field of geographical data.

SOAP, Simple Object Access Protocol
SOAP is a communication model between distant programs, often on the Web.

SOLAP, spatial online analytical processing.

SVG, http://www.w3.org/Graphics/SVG/
SVG, Scaleable Vector Graphics, is an XML format to represent 2D graphics with geometries.

TOMCAT or APACHE TOMCAT, http://tomcat.apache.org/
TOMCAT is an open source piece of software that can host programs so that they can be queried on the Web.

UDIG, http://udig.refractions.net/
UDIG is an open source piece of GIS software. One of its specificities is to be based on the Eclipse technology (Eclipse rich client). This implies a great code flexibility.

UML, Unified Modeling Language.

XERCES, http://xerces.apache.org/
XERCES is an open source project to read XML flows. It offers libraries in different languages: JAVA, PERL, C++.

XML, Extensible Markup Language.

XSLT, Extensible Stylesheet Language Transformations.

XMI, http://www.omg.org/spec/XMI/
XMI, XML Metadata Interchange, is a standard exchange format based on XML allowing us to write models.

WKB, Well-Known Binary, is a standard geometry encoding model in Binary Large Objects (called BLOB).

WKT, Well-Known Text, is a standard geometry encoding model in text.

List of Authors

Olivier BEDEL
Société Alkante
Cesson Sévigné
France

Erwan BOCHER
IRSTV FR CNRS 2488
Ecole Centrale de Nantes
France

Mickaël BRASEBIN
IGN-COGIT
University of Paris Est
Saint-Mandé
France

Bénédicte BUCHER
IGN-COGIT
University of Paris Est
Saint-Mandé
France

Paule-Annick DAVOINE
Grenoble-INP
Laboratoire LIG
Saint-Martin-d'Hères
France

Jean-Christophe DESCONNETS
IRD
Espace-Dev
Maison de la Télédétection
Montpellier
France

Sébastien FERRÉ
IRISA
University of Rennes 1
France

Julien GAFFURI
Joint Research Centre
European Commission
Ispra
Italy

Jérôme GENSEL
University Pierre
Mendès France
Laboratoire LIG
Saint-Martin-d'Hères
France

Éric GROSSO
IGN-COGIT
University of Paris Est
Saint-Mandé
France

Florence LE BER
LHYGES
LORIA
Ecole Nationale du Génie de l'Eau
et de l'Environnement de Strasbourg
France

Thérèse LIBOUREL
Espace-Dev
University of Montpellier 2
France

Jean-François MARI
LORIA
University of Nancy 2
France

Alina MIRON
University Joseph Fourier
Laboratoire LIG
Saint-Martin-d'Hères
France

Bogdan MOISUC
LIG
University Joseph Fourier
Saint-Martin-d'Hères
France

Moritz NEUN
Institut de Géographie
Zurich
Switzerland

Julien PERRET
IGN-COGIT
University of Paris Est
Saint-Mandé
France

Gwendall PETIT
IRSTV FR CNRS 2488
Ecole Centrale de Nantes
France

Nicolas REGNAULD
Ordnance Survey
Southampton
UK

Olivier RIDOUX
IRISA
University of Rennes 1
France

Marlène VILLANOVA-OLIVER
LIG
University Pierre Mendès France
Saint-Martin-d'Hères
France

Robert WEIBEL
Institut de Géographie
University of Zurich
Switzerland

Index

3D data, 79, 81

C

cartographic
　generalization, 257-259, 279
　representation, 7, 36, 147
　view, 156, 159, 160, 166, 175, 176
cloud computing, 11, 168, 270, 272
cropping pattern, 194, 198, 203, 204

D

data acquisition, 16, 25, 97, 308
digital terrain model (DTM), 51, 54, 81
drainage basin, 51, 52, 55, 56

E

ESRI, 10, 229
ESRI *shapefile*, 6, 37, 70, 72-74, 81, 295

F

field pattern, 190, 194, 195, 201-204, 210, 290
Food and Agriculture Organization (FAO), 220, 231

G

GENEXP-LANDSITES, 11, 189-210, 286, 288-292, 301
GENGHIS, 7, 9, 93, 116, 117, 121-149, 285-289, 291, 312
geocoding, 57, 59
geolocation, 48, 56-61

geomatics, 1, 2, 5, 9-11, 13-15, 17, 18, 62, 69, 115, 224, 287-289, 293-295, 297, 302, 303, 305, 307, 311-313
geomatics software, 2-12, 294, 303
Gibbs process, 192, 199
GIS software, 5-7, 10, 11, 18, 27, 28, 67, 70
GIS-tool, 5-9, 11, 29

I

International Organization for Standardization ISO), 5, 6, 14, 1n 67, 69, 74, 77, 81, 218, 219, 226, 227, 228, 231, 232, 237, 240, 241, 252, 292, 295-299
IRSTV, 25-27, 28, 29, 31, 32, 62, 64, 288

M

MAGIS, 9, 15, 86, 307, 308
map generalization, 259, 260
Markov model (MM)
 hidden Markov models (HMM), 198, 199, 200
MDWEB, 6, 215-253, 285-290, 292, 296, 312
MeigeVille, 27, 29
metadata
 exchange, 236-238
 model, 219, 226, 249, 252, 306

profile, 224, 226, 227, 228-230, 234, 237, 240, 246
sheet, 228, 233-235, 238
metamodel, 77, 100-102, 112, 295
monolingual, 241, 242
multilingual, 231, 241, 242, 244

N

natural risk, 108, 109, 115, 116, 122, 123, 141-146, 149, 253
natural risk modeling, 115

O

Open Geospatial Consortium (OGC), 5-7, 14, 28, 30, 31, 35, 38, 67, 69, 87, 107, 131, 169, 218, 219, 238, 241, 246, 248, 249, 252, 260, 261, 265, 292, 295-299
open source, 14, 26, 62, 64, 68, 70, 78, 87, 216, 220, 251, 252, 257, 261, 264, 266, 277, 291, 295, 301

P

Perval, 57, 58, 59, 60, 61
planar graph, 74, 75
pooling, 2, 4, 5, 7, 11-18, 290, 293, 294, 298, 299, 301-305, 307, 311, 312
post-production, 194, 200, 201, 206

R

random seeds, 202, 203, 208
raster, 29-35, 39, 42, 43, 48, 51, 54, 55, 124, 132, 190, 229, 289, 299
rectification, 57, 59, 60, 61
resource cataloging, 219, 225, 228, 230, 233
reuse, 5, 14, 18, 68, 87, 235, 267, 268, 272, 292, 294-296, 298, 302, 305, 307
risk management, 1, 123, 215, 250, 253

S

Simple Features SQL (SFS), 29-32, 34, 35, 53
Simple Knowledge Organization System (SKOS), 231, 243, 244
spatial
 hydrology, 29, 51, 62
 information, 9, 26, 193
spatial data infrastructure (SDI), 27, 28f, 29, 62, 216, 244, 250
spatiotemporal
 information systems (STIS), 9, 93, 97, 99, 121, 122, 125, 130, 140, 141, 285

T

territory management, 148, 215
tessellation
 rectangular, 194, 195, 196, 197, 208
 space, 194, 285, 292
 Voronoï, 202, 207

U

UML
 class diagrams, 92
 formalism, 227, 228
urban
 data, 25, 27
 management, 26, 27
 model, 27, 306
 sprawl, 48
urbanized surface, 48-51, 52, 53, 56
user community, 26, 85, 86, 93, 115, 122, 146-148, 184, 221, 226, 249-251

V

vector, 3, 29-31, 34, 37, 39, 48, 132, 141, 228, 229, 231, 239, 262, 289, 290
Voronoï diagram, 194-196, 200, 201, 203, 204